ADVANCED PROCESSES AND TECHNOLOGIES FOR ENHANCED ANAEROBIC DIGESTION

MOST RECENT ADVANCES IN ANAEROBIC DIGESTION INSIDE ONE DOCUMENT!

Luxmy Begum, Ph. D., P. Eng., PMP

A Green Nook Press Publication
Toronto, Ontario, Canada.

Authored by Luxmy Begum, Ph. D., P. Eng., PMP
Website: www.TheEcoAmbassador.Com

First Edition, 2014
ISBN (Book): 978-0-9939045-0-9
ISBN (Electronic Book Text): 978-0-9939045-1-6

A Green Nook Press Publication
Toronto, Ontario, Canada.

Ordering Information:
Hard Copy: Amazon.com
Electronic Copy (PDF): http://www.theecoambassador.com/EnhancedAnaerobicDigestion.html

About the Author:

Dr. Luxmy Begum is an Environmental Engineer with over 15 years of process, research and project management experiences in sustainable technologies and renewable energy sector. Dr. Begum received a bachelor degree in civil engineering from the Indian Institute of Technology (IIT), Chennai, India; a master degree in environmental engineering from the Asian Institute of Technology (AIT), Bangkok, Thailand and a Ph.D. Degree in Environmental Engineering from the University of Tokyo, Japan. Dr. Begum was also a recipient of the Canadian Government NSERC Post Doctoral Research Fellowship at Environment Canada on alternative energy production (Biogas and Hydrogen). As part of her professional experience, Dr. Begum had worked with some major environmental engineering firms and served as a consultant both nationally and internationally. Dr. Begum's professional experience includes innovative process solution, plant design and upgrade, technology selection and evaluation, equipment selection and procurements as well as project management and execution. She has successfully managed and delivered various environmental and renewable energy projects throughout North America for municipal, provincial, and private sector clients. Dr. Begum has authored over fifty (50) international journals, conference papers and technical reports. This book is a result of her extensive research as well as professional experience on the topic.

Preface

Anaerobic digestion technology has made some significant advancement in recent years. Different government agencies and professional groups have published several reports time to time on different individual steps of the process– like feed pretreatment, biogas upgrading or resource recovery of digestate etc., covering only some of these advancements. In addition, there are several hand books available that although cover the whole process, but only in a very basic level. There remains a lack of a comprehensive document that captures all the promising advancements and emerging technologies for the entire anaerobic digestion treatment process; starting from the beginning to the end. I have authored this book to fulfill that need in the industry. This book covers almost all the recent advancements in the industry starting from the emerging feed stocks to resource recovery of digestate; providing sufficient details on theories and practices, technologies and technology suppliers. I am confident that both the beginners and the experienced professionals will find this book useful.

Luxmy Begum Ph. D., P. Eng., PMP
Ontario, Canada

Table of Contents

INTRODUCTION 1

BASIC ANAEROBIC DIGESTION PROCESS OVERVIEW 3

ANAEROBIC DIGESTION PROCESS OVERVIEW 3
  Hydrolysis 3
  Fermentation 3
  Methanogenesis 3
ANAEROBIC DIGESTION TECHNOLOGY OVERVIEW 5

ADVANCEMENT IN FEED COMPOSITION 7

CO-DIGESTION OF FEEDSTOCKS 7

CONVENTIONAL AND EMERGING FEEDSTOCKS FOR ANAEROBIC DIGESTION 8
  Feedstock Type 8
  Main Characteristics of Anaerobic digestion Feedstocks 11
  Energy Value of Different Feedstocks 13
  Feedstock Biogas Potential - BMP Assay 15

FEED PRE-TREATMENT FOR ENHANCE ANAEROBIC DIGESTION 18

PRE-TREATMENT CATEGORIES AND RELATED TECHNOLOGIES 18
  Sorting, Separation and Homogenization 19
  Pasteurization and Pathogen Reduction 21
  Hydrolysis Enhancement and Process Improvement 21
PRE-TREATMENT OF FOOD WASTE AND ORGANIC FRACTION OF MUNICIPAL SOLID WASTE 26
  Example of Food Waste/ OFMSW Pre-treatment Plants and Technology Suppliers 26

PROCESS CONSIDERATION, CONFIGURATION AND TECHNOLOGY SELECTION FOR ENHANCED ANAEROBIC DIGESTION 30

PROCESS CONSIDERATIONS FOR OPTIMAL PERFORMANCE 30

PROCESS SELECTION AND CONFIGURATIONS FOR ENHANCE PERFORMANCE 34
  Process Selection Consideration - Feedstock's Solid Content 34
  Process Selection Consideration - Digester's Operating Temperature 37
  Process Selection Consideration - Digester's Stages and Rector Configuration 38
  Process Selection Consideration - Digester's Feeding Mode 42
  Process Selection Consideration – Enhancing Solids Loading and Digestate Recirculation ... 42
  Different Digester Types with Some of Their Operating Characteristics 44

CO-DIGESTION PROCESS CONSIDERATION 47

    Feedstock Reception, Pre-Treatment and Feeding.. 47
    Co-digestion Operating Conditions and Challenges 48

BIOGAS CLEANING, UPGRADING AND UTILIZATION **53**

COMPOSITION OF BIOGAS 53

BIOGAS CLEANING TECHNOLOGIES 54

BIOGAS UPGRADING TECHNOLOGIES 57

    Absorption/Scrubbing 57
    Pressure Swing Adsorption.. 60
    Membrane Separation 61
    Cryogenic Separation/Distillation 62
    Comparison of Different Upgrading Technologies 63
    Biogas Upgrading Plant/Technology Suppliers 64

BIOGAS UTILIZATION 67

DIGESTATE TREATMENT AND RESOURCE RECOVERY **69**

DIGESTATE TREATMENT AND ENHANCEMENT OPTIONS 69

    Physical Treatment and Enhancement Options 70
    Thermal Treatment and Enhancement Options.. 71
    Conversion Treatment and Enhancement Options 73
    Biological Treatment and Enhancement Options.. 76
    Chemical Treatment and Enhancement Options 79

RESOURCE RECOVERY FROM DIGESTATE 83

REFERENCES **86**

List of Figures

FIGURE 2.1: ANAEROBIC PROCESS SCHEMATIC 4

FIGURE 2.2: AN ADVANCED ANAEROBIC DIGESTION PLANT 5

FIGURE 2.3: A SIMPLIFIED ANAEROBIC DIGESTION TECHNOLOGY OVERVIEW 6

FIGURE 3.1: ENERGY VALUE PROFILE OF DIFFERENT ANAEROBIC DIGESTION FEEDSTOCKS...............13

FIGURE 3.2: BMP VIAL 15

FIGURE 5.1: A STAGED MESOPHILIC DIGESTION CONFIGURATION 39

FIGURE 5.2: TWO ACID/GAS (AG) PHASED DIGESTION CONFIGURATIONS...................................40

FIGURE 5.3: A TEMPERATURE PHASED ANAEROBIC DIGESTION (TPAD) CONFIGURATION 41

FIGURE 5.4: A STAGED THERMOPHILIC DIGESTION CONFIGURATION 41

FIGURE 5.5: AN EXTENDED SOLIDS RETENTION (ESR) DIGESTION CONFIGURATION........................43

FIGURE 5.6: A WASTE RECEPTION HALL..47

FIGURE 5.7: A FOOD WASTE SEPARATION SYSTEM IN A WASTE TRANSFER FACILITY 48

FIGURE 6.1: A SCHEMATIC DIAGRAM OF WATER SCRUBBING..58

FIGURE 6.2: A SCHEMATIC DIAGRAM OF ORGANIC PHYSICAL SCRUBBING 59

FIGURE 6.3: A SCHEMATIC DIAGRAM OF PRESSURE SWING ADSORPTION 60

FIGURE 6.4: A SCHEMATIC DIAGRAM OF TWO STAGE MEMBRANE SEPARATIONS 61

FIGURE 6.5: SCHEMATIC DIAGRAM OF A CRYOGENIC SEPARATION..62

FIGURE 6.6: INJECTION SYSTEM USED TO PUMP RNG INTO A NATURAL GAS LINE AT A FARM-BASED BIOGAS SYSTEM 67

FIGURE 6.7: SOME BENEFICIAL END USE OF BIOGAS ALONG WITH THEIR CLEANING AND UPGRADING REQUIREMENTS. 68

FIGURE 7.1: HUBER SOLAR ACTIVE DRYER SRT 72

FIGURE 7.2: THE J-VAP® DEWATERING AND DRYING SYSTEM 73

FIGURE 7.3: A GASIFICATION SYSTEM BY NEXTERRA 74

FIGURE 7.4: A PYROLYSIS SYSTEM BY 3R AGROCARBON..75

FIGURE 7.5: MULTI BIN COMPOSTING, IN-VESSEL COMPOSTING AND HOOP-STRUCTURE COMPOSTING 76

FIGURE 7.6: A REED BED SYSTEM 77

FIGURE 7.7: A FULL SCALE MEMBRANE BASED SYSTEM TO TREAT DIGESTATE BY BKT21 IN NETHERLANDS........78

FIGURE 7.8: A VIEW OF AN ALGAL BIOREACTOR FOR BIOFUEL PRODUCTION.................................78

FIGURE 7.9: STRUVITE RECOVERY AT DURHAM WASTEWATER TREATMENT PLANT, TIGARD, OREGON, USA........80

FIGURE 7.10: VARIOUS RESOURCE RECOVERY OPTIONS FROM DIGESTATE 85

List of Tables

TABLE 3.1: DIFFERENT ANAEROBIC DIGESTION FEEDSTOCK AND THEIR SOURCES 10

TABLE 3.2: COMMONLY USED FEEDSTOCKS AND SOME OF THEIR IMPORTANT CHARACTERISTICS 12

TABLE 3.3: APPROXIMATE BIOGAS YIELDS OF VARIOUS CO-DIGESTION FEEDSTOCKS 14

TABLE 5.1: EXAMPLE OF FACILITIES WITH MULTI-STAGE ANAEROBIC DIGESTERS 43

TABLE 5.2: COMMON DIGESTER TYPES WITH SOME OF THEIR OPERATING CHARACTERISTICS 44

TABLE 5.3: OPTIMUM BLEND RATIO OF SOME CO-DIGESTION FEEDSTOCKS 49

TABLE 6.1: BIOGAS CONTAMINANTS AND RELATED CLEANING TECHNOLOGIES 54

TABLE 6.2: COMPARISON OF DIFFERENT UPGRADING TECHNOLOGIES 63

TABLE 6.3: LIST OF DIFFERENT BIOGAS UPGRADING PLANT/TECHNOLOGY SUPPLIERS 64

TABLE 7.1: A SUMMARY OF DIFFERENT TREATMENT AND ENHANCEMENT OPTION FOR DIGESTATE. 82

Chapter 1

Introduction

Anaerobic Digestion is an almost 200 years old technology that has evolved over the years. In Anaerobic Digestion (AD) process, organic materials get decomposed and broken down by anaerobic microorganisms in the absence of oxygen. The main product of anaerobic digestion is biogas, which is an alternative and renewable energy. Biogas typically consists of 60% of Methane and 40% of carbon dioxide. Increasing demand for energy and scarcity of fossil fuel have created a renewed interest in alternative energy like biogas in recent years. In addition, biogas can be generated utilizing various organic products like food waste, plant and yard waste, oil and grease etc., that are considered waste. This waste to energy conversion has open up new economic and recycling options.

Anaerobic Digestion technology can provide a complete solution to organic waste management. Biogas generated through the process can be converted to electricity or heat through Combined Heat and Power (CHP) generation unit. Biogas can also be refined further to pipeline quality gas (comparable to natural gas, which is 97% methane) and used as fuel to run public transportation or waste collection vehicles. Thus producing biogas from local organic waste can provide a stable, local supply of renewable energy. In addition, the nutrients and solid materials from the digestate (slurry output from digester) can be concentrated and converted to valuable resources such as compost or dry organic fertilizer. Thus, with anaerobic digestion of organic waste, it will be possible to offset a significant amount of fossil fuel consumption while keeping organic waste out of the landfill and avoiding associated greenhouse emissions. So, there are two very promising advantages of using anaerobic digestion Technology:

- Anaerobic digestion can provide an alternative energy source which is promising in the midst of energy scarcity and rising oil price.

- Anaerobic digestion can convert waste to bioenergy, renewable fuels and valuable resources and thus divert waste from landfill and reduce green house gas emission.

With renewed interest in biogas, anaerobic digestion technology has also advanced rapidly in recent years. An advanced, high performance process not only improves processing time and cost but also the quality and quantity of final products. There are several reasons for using high performance anaerobic digesters and auxiliary technologies. Some of them are:

- Increased Biogas Production

- Optimized Resource Recovery

- Improved Biosolid Quality

- Enhanced Energy Conversion

- Increased Digester Capacity

- Enhanced Operating Characteristics

- Faster Return on Investment

- Reduced Residual

With the aim of maximizing biogas production, different feed compositions and process configurations were tried in the industry. Some advancement have also been made to solve existing operational issues and limitations like improving digesters mixing, enhancing feeding and withdrawing, eliminating foaming issues or simply increasing digestion capacity of existing digesters. Some of these advancements in process, technologies as well as feed composition have produced very promising and outstanding results.

The purpose of this book is to summarize these recent advancements in the anaerobic digestion process and auxiliary technologies for the greater benefit of the industry.

Chapter 2

Basic Anaerobic Digestion Process Overview

Anaerobic Digestion Process Overview

Anaerobic digestion is a collection of natural and biological processes in which anaerobic microorganism break down organic materials in the absence of oxygen. Biogas is produced when organic materials get decomposed and broken down. The whole degradation and biogas production process can be divided into three basic steps:

Hydrolysis

Hydrolysis is the first step where particulate materials get converted to soluble compounds suitable for further breakdown in the next step. In this step, organic material transformed into simple liquefied monomers and polymers like glucose, fatty acids, amino acids etc.

Fermentation

The second step of degradation is known as fermentation or acidogenesis where the products of hydrolysis such as simple sugars, amino acids, fatty acids etc. break down further and produce final products of fermentation such as hydrogen (H_2), carbon dioxide (CO_2) and acetate.

Methanogenesis

The third and final step of anaerobic degradation is known as methanogenesis. In this process, one group of microorganism known as *aceticlastic methanogens,* converts acetate into H_2 and CO_2. Then a second group of microorganism referred to as Hydrogen-utilizing methanogens combines H_2 and CO_2 into methane (CH_4). The end product of methanogenesis is biogas, a mixture of methane and carbon dioxide.

Figure 2.1 presents a schematic diagram of anaerobic digestion process overview.

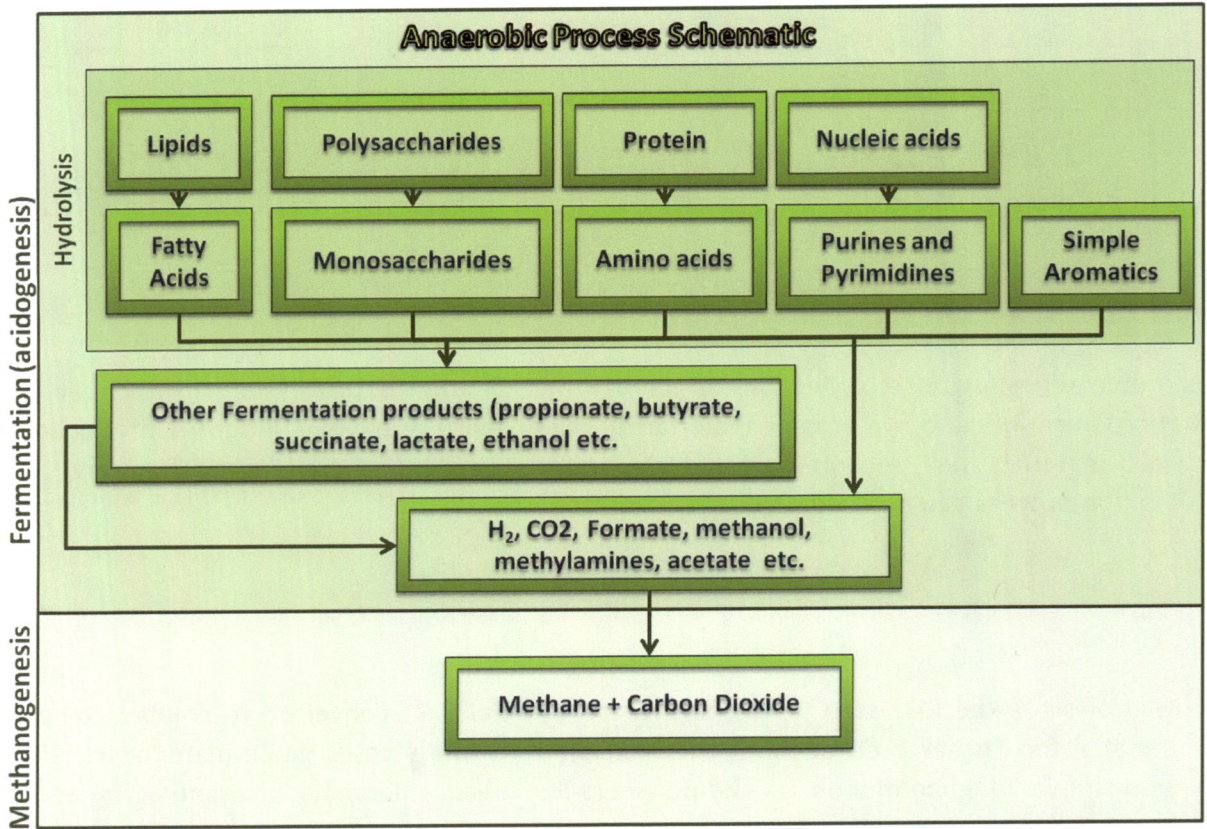

Figure 2.1: Anaerobic Process Schematic
(Data Source: Metcalf and Eddy, 2003)

Anaerobic Digestion Technology Overview

In Anaerobic Digestion (AD) technology, anaerobic process mechanisms and naturally occurring anaerobic microorganisms are utilized to treat organic wastes (like manure, sewage sludge, food waste or other biodegradable wastes). In the engineered process, an optimum environment is created for various microorganisms to break down different organic materials in the absence of oxygen. In general, anaerobic digestion is designed in an enclosed vessel known as anaerobic digester. The process can be performed with or without mixing and either in batch, semi-continuous or continuous mode.

Figure 2.2: An Advanced Anaerobic Digestion Plant

The main product of anaerobic digestion is biogas, which is an alternative and renewable energy. Biogas typically consists of 60% of methane and 40% of carbon dioxide. It also contains trace amount of other gases like nitrogen, hydrogen and hydrogen sulphide.

The effluent of digester, known as digestate is a combination of residual solids and water. Further resources can be recovered from digestate or its fractions (fibre and liquor) and can be utilized as compost or fertilizer.

There are four basic steps involved in anaerobic digestion treatment technology:

1. Feedstock receiving and handling
2. Anaerobic digestion of feed stocks
3. Biogas capture, utilization and upgrade if desired
4. Digestate handling and resource recovery

A simplified overview of anaerobic digestion technology is shown in the Figure 2.3. The details of the recent advancements and technologies evolved in these four steps will be discussed in the later chapters.

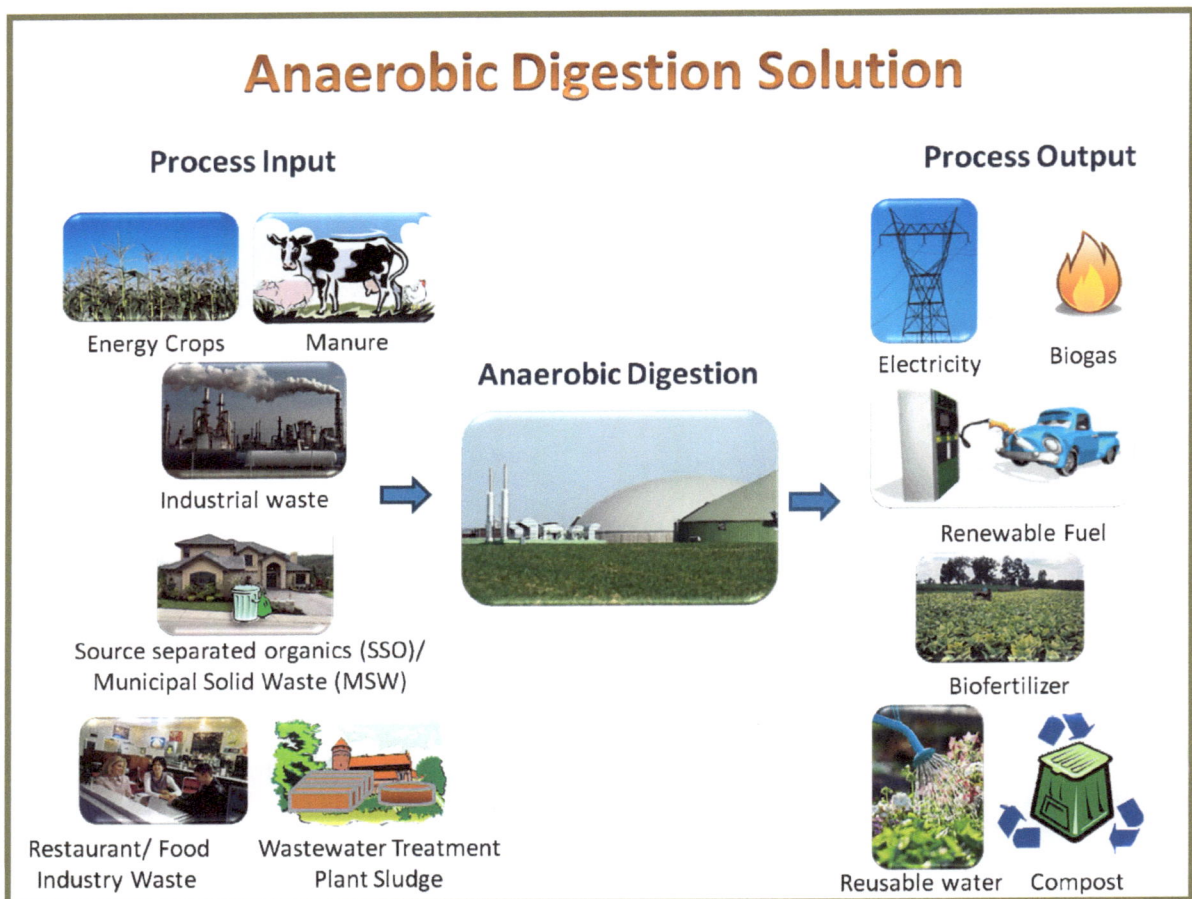

Figure 2.3: A Simplified Anaerobic Digestion Technology Overview

Chapter 3

Advancement in Feed Composition

Co-digestion of Feedstocks

Anaerobic digestion process was originally invented to treat manures and sewage sludge. But manures and sewage sludge are not the best quality feedstocks as they had already gone through a digestion process inside the body where majority of their energy contents had been consumed. So in order to enhance the gas production, focus has been shifted to other high energy feedstocks that can be blended with manures or sewage sludge and digest together. This process of blending and digesting two or more feedstock at the same time in an anaerobic digestion process is known as co-digestion.

The most common situation of co-digestion is when a higher amount of basic feedstock like sewage sludge or manure is mixed and digested together with a lower amount of high strength organic waste like FOG (Fats, Oils and Grease) or food waste. So, the primary feedstock in a digester may be manures and sewage sludge. Then one or more type of energy rich feedstock such as corn silage, food processing waste, slaughterhouse waste, FOG, organic fraction of household waste/municipal solid waste etc. can be added to the digester and a higher methane and biogas production can be achieved.

Among the energy rich feedstocks FOG, food waste and organic fraction of the municipal waste stand out because of their high gas production and energy recovery potential. Food waste has three times higher methane production potential when compared to biosolid and also much higher biodegradability than biosolid (86-90%), as per USEPA. It means, even when a higher amount of food waste are added to the digester as a feedstock, the final residue will only increase by a smaller amount but the gas production will increase significantly. The food waste or organic waste can be combined-waste collected from restaurants, school cafeterias, produce markets, and fish markets, house hold kitchen waste and food leftovers, as well as yard waste.

Rather than sending these energy-rich wastes to land fill they can be co-digested in an anaerobic digester to reap some of the potential benefits of co-digestion as follows:

Renewable Energy Generation: Due to Increase demand and scarcity of fossil fuels, increasing renewable energy production is more crucial. Co-digestion of food waste and organic waste allows increased biogas production that can be utilized in the plant itself and also can be sold to the grid when in excess.

Greenhouse Gas Reduction: When food wastes or organic fractions of municipal solid wastes (OFMSW) are dumped in landfill, they generate a large amount of methane, a known green house gas that usually escapes in the atmosphere. But if these same wastes are diverted from landfill and co- digested in anaerobic digesters, almost three times higher methane production (US EPA) can be achieved compared to biosolid, which of course can be utilized further.

Waste to Energy: Most cities/municipalities are looking for ways to reduce waste to land fill and maximize recycling. Utilization of food waste and OFMSW as anaerobic digestion feedstocks not only minimizes waste streams but also allows the recovery of valuable energy and resources from them.

Cost Recovery: The anaerobic digestion plant can recover some of its operational cost from the enhanced gas production/onsite power generation. They can also be benefitted from tipping fees for accepting food waste.

Conventional and Emerging Feedstocks for Anaerobic Digestion

Feedstock Type

Traditionally manure, animal waste and waste water sludge were the main feedstocks for anaerobic digester. But recently food processing waste and organic waste from household have drawn significant attention as promising feed stocks that can increase gas production significantly. In addition, agricultural residues, yard waste and even municipal solid waste can also be used as feedstocks for anaerobic digestion after appropriate pre-treatments. Introduction of new pre-treatment technologies and innovative digester types have opened the option of recovering energy from diverse waste streams rather than sending them to land fill.

Some common and emerging feedstocks for anaerobic digestion are discussed below:

Manure

Solid or liquid manure produced in animal farms (cow, pig, chicken etc.) is one of the main feedstocks for biogas production.

Domestic Source Separated Organic (SSO) Waste

This waste category include separately collected kitchen waste like fruit and vegetable waste, food waste, coffee and tea filters and other organic leftovers.

Yard and Plant Waste

Yard or plant waste include leaves; garden wastes likes flowers, plants, vegetables; branches and tree trimmings, grass clippings etc.

Energy Crops

Energy crops are plants suitable for producing biogas and other renewable energy. Examples of energy crop include corn, Sudan grass, maize, millet, white sweet clover, sugar cane etc.

Sewage Sludge

Almost all Wastewater treatment plants produced sewage sludge and this sewage sludge can also be used as feedstocks for anaerobic digestion to produce biogas. Although due to its limited potential for biogas production it is not treated as a high quality substrate.

Industrial and Restaurant Food Waste

Industrial food wastes mainly come from food and meat production sites like slaughter house waste, whey, potato mash, brewer grains etc. In addition food waste can be collected from restaurants, grocery stores, and other food handling facilities and used as high potential biogas producing substrates.

Potential Co-digestion Feedstocks

Many high strength and emerging feedstocks have the potential to be a co-digestion feedstocks. A good co-digestion feedstock should fulfill certain requirements such as:

1. should be easily biodegradable, with a high biogas production potential,
2. Should not contain any inhibitory substances for the process
3. Should have a macro/micro nutrients that may enrich the primary feedstocks
4. should be available in sufficient quantities and at a reasonable price
5. Should have a easy reception and storage option

The lists of such feedstocks are as follows:

- Fats, oils, and grease (FOG)
- Source Separated organics (SSO)
- Restaurant food scraps
- Food manufacturing and Baking industry wastes
- Breweries and distilleries
- Slaughter house wastes
- Dairy wastes

A list of potential anaerobic digestion feedstocks and their sources is provided in Table 3.1.

Table 3.1: Different Anaerobic Digestion Feedstock and their Sources

Agricultural and Farm	Municipal	Industrial
Harvest Remains Energy Crops like corn, Sudan grass, maize, millet, white sweet clover etc. Algal Biomass Manure (Pig, cow, poultry)	Source Separated organics (SSO) like fruit and vegetable waste, food waste, coffee and tea filters and other organic leftovers. Yard or garden waste like leaves, plants, vegetables, branches and tree trimmings, grass clippings etc. Sewage Sludge Municipal Solid waste (MSW)	food/beverage processing industry waste Dairy Industry waste Starch industry waste Sugar industry waste Slaughterhouse waste

Main Characteristics of Anaerobic digestion Feedstocks

Anaerobic digestion feedstocks can vary widely based on their organic fraction, water content and biodegradability. Usually, it is expected that organic content of the feedstocks should be in the range of 70% – 95% of the dry matter content. In general, feedstocks with less than 60% dry matter organic content are considered not so worthwhile for anaerobic digestion.

The nutrient ratio, specially the Carbon to Nitrogen Ratio (C:N) is very important for the microbes in biodegradation process. The ideal carbon to nitrogen (C:N) ratio for anaerobic digestion ranges from approximately 20:1 to 30:1 (AgSTAR/US EPA). The optimum carbon, nitrogen and phosphorus (C:N:P) ratio for anaerobic digestion has been reported to be 100:5:1 (Feedstocks for Anaerobic Digestion, Steffen's et al., 1998; www.adnett.org/dl_feedstocks.pdf).

The water content of feedstocks is also very important for anaerobic digestion. The high water content feedstocks require higher digester volume and higher heat input; thus increase the process cost. On the other hand, a higher solid content feedstock has mixing issue, solid settling issue as well as clogging and scum layer formation. That is why the optimum total solid concentration for a completely mixed anaerobic digester is suggested in the range of 6 to10% (Feedstocks for Anaerobic Digestion, Steffen's et al., 1998; www.adnett.org/dl_feedstocks.pdf).

Some commonly used feedstocks and some of their important characteristics are listed in Table 3.2.

Table 3.2: Commonly Used Feedstocks and Some of their Important Characteristics

Feedstock	Total Solids TS (%)	Volatile Solids (% of TS)	C:N Ratio
Pig Slurry	3–8 *	70–80	3–10
Cow Slurry	5–12*	75–85	6–20**
Chicken Slurry	10–30	70–80	3–10
Whey	1–5	80–95	N/A
Ferment, Slops	1–5	80–95	4–10
Leaves	80	90	30–80
Wood Shavings	80	95	511
Straw	70	90	90
Wood Wastes	60–70	99.6	723
Garden Wastes	60–70	90	100–150
Grass	20–25	90	12–25
Grass Silage	15–25	90	10–25
Fruit Wastes	15–20	75	35
Food Remains	10	80	N/A

* – Depending on dilution
**– Depending on straw addition
Data Source: Steffen's et al., (1998); Feedstocks for Anaerobic Digestion,
www.adnett.org/dl_feedstocks.pdf

Energy Value of Different Feedstocks

Different feedstocks have different methane production potential or different energy recovery value in anaerobic digestion. In addition, process selection, reactor design and operation are also done based on input feedstocks' characteristics. Moreover, quantity and quality of end product and process economics are also functions of feedstock characteristics. That is why feedstock selection and optimum feedstock mix are crucial for anaerobic digestion process design.

Anaerobic digestion process was originally designed for manures and sewage sludge. But these two feedstocks do not have much energy value left as they have already gone through some digestion inside the animal's body.

In general, the feedstock of high organic content that is easily degradable by anaerobic bacteria are considered good feedstock for anaerobic digestion; such as food and food processing waste, fats, oil and grease etc. Some energy crops also have high level of biogas production potential such as corn, maize and grass silage. On the other hand poor candidates for anaerobic digestion are yard waste, woody waste etc. as most anaerobes are unable to degrade lignin, the major fraction of these wastes.

Figure 3.1 shows energy value profile of different anaerobic digestion feedstocks:

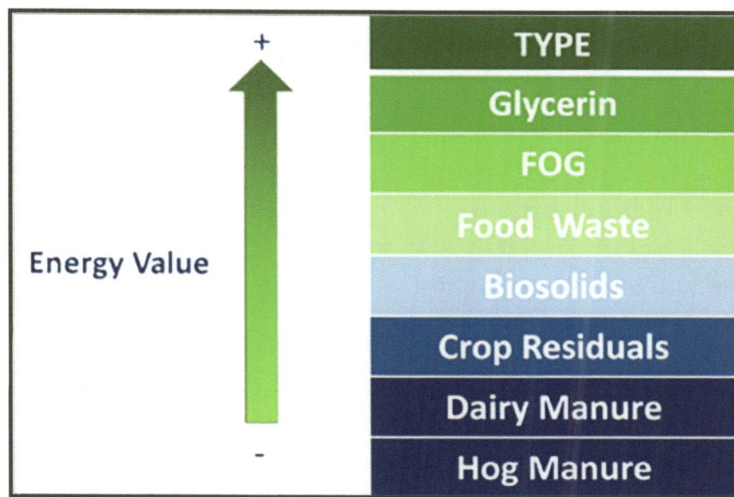

Figure 3.1: Energy Value Profile of Different Anaerobic Digestion Feedstocks
(Source: Clemens Halene, Quasar Energy Group, Feedstock Streams for Anaerobic Digesters; as reported in Richardson, 2010)

Table 3.3 shows various co-digestion feedstocks and their approximate biogas yields in cubic meter (m³) per ton of organic solids.

Table 3.3: Approximate Biogas Yields of Various Co-digestion Feedstocks

Material	Biogas Yield (m³/Ton of Organic Solids)
Harvest Residues: Straw, stems, sugar beet toppings, fibrous material	375
Animal Manures	200–500
Food industry waste: Dough, confectionary waste, whey	400–600
Yeast And Yeast Like Products: Yeast and sludge from breweries, wine making, distilleries	400–800
Residues From Animal Feed Production: Expired feed	500–650
Slaughterhouse Waste: Flotation sludge, animal fat, stomach and gut content, blood	550–1,000
Wastes From Plant and Animal Fat Production: plant oil, oil seed, fat, bleaching earth	1,000
Pharmaceutical Wastes: Proteinacious wastes, bacterial cells and fungal mycelium	1,000–1,300
Waste From Pulp And Paper Industry	400–800
Sludge from gelatine and starch production	700–900
Biowastes From Source Separated Collection	400–500
Market Waste	500–600
Sewage Sludge	250–350
Data Source: Braun, R. and Wellinger, A.; Potential of Co-digestion by IEA Bioenergy, Task-37 (http://www.iea-biogas.net/)	

Feedstock Biogas Potential - BMP Assay

Biogas generating potential of any feedstock or feedstock blend can be predicted through a lab based approach, commonly known as Biochemical Methane Potential (BMP) Assay.

BMP assay is an efficient and economic method for estimating of biomass conversion and biogas yield potential of any feedstock or feedstocks blend. It was first established by Owen et al. (1979). In that scientific paper they explain the BMP assay method as well as how result can be explained and used.

In general, The BMP assay is conducted by combining a single feedstock or feedstocks blend, inoculums, and stock solutions in a closed assay bottle for a batch system mode. The bottle should be flashed with nitrogen gas to maintain anaerobic environment and closed with a septum for the convenience of gas collection via syringe. Inoculums are an active anaerobic culture added to the bottle to reduce any lag time and initiate the digestion process immediately. Stock solutions are mixture of macronutrients, micronutrients and vitamins to assure that biogas production are not limited during the assay. A water bath can be used for temperature control and a shaker can be used for mixing the solutions. BMP Assay can be continued for as long as 60 days and biogas produced can be measured periodically or continuously.

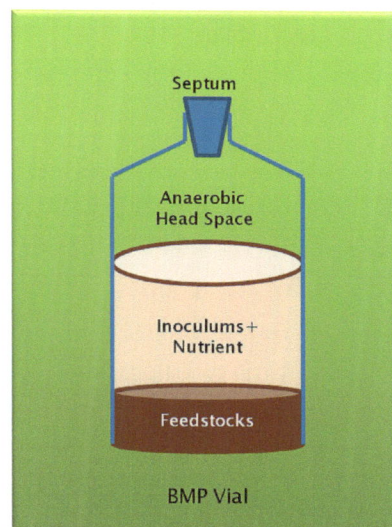

Figure 3.2: BMP Vial

A control solution bottle containing inoculums and stock solutions should also be used to determine gas production from the inoculums alone. BMP assay should always be conducted in replication to assure the repeatability of the test results.

The following parameters are usually analyzed in a BMP assay to evaluate the feedstock or feedstock blend characteristics:

- **Before and after feedstock characteristics:** The feedstock characteristics like pH, chemical oxygen demand (COD), total solids (TS), and volatile solids (VS) are checked before and after the BMP assays. The before assay results help to determine the feedstock quantity required for maintaining the assay for as long as 30 to 60 days. The after assay feedstock characteristics reveal organic material destruction and biodegradability of the feedstocks during the assay.

- **Total biogas production and gas composition:** Total gas production can be measured during the BMP test either by collecting it manually or by continuously using commercially available gas meters and softwares. Later biogas composition can be analyzed by a gas chromatograph that is set to determine the concentrations of methane, carbon dioxide, nitrogen, and hydrogen sulfide gases.

It is suggested that BMP results should also be verified at a bench and pilot scale level. It is strongly recommended that full scale designs should not be based on BMP results as full scale digesters are often run at continuous mode while BMP tests are at batch mode (http://www.extension.org/pages/26617/feedstocks-for-biogas).

Some additional BMP methods for reference are listed below:

- ASTM Standard E2170 – 01 (2008), Standard Test Method for Determining Anaerobic Biodegradation Potential of Organic Chemicals Under Methanogenic Conditions. ASTM International, West Conshohocken, PA.

- ISO 11734:1995 Water quality – Evaluation of the "ultimate" anaerobic biodegradability of organic compounds in digested sludge – Method by measurement of the biogas production. International Standards Organization.

- Moody, L., R. Burns, W. Wu-Haan, and R. Spajić. (2009). Use of biochemical methane potential (BMP) assays for predicting and enhancing anaerobic digester performance. In Proceedings of the 4th International and 44th Croatian Symposium of Agriculture. Optija, Croatia.

- Gunaseelan, V.N. (1997). Anaerobic digestion of biomass for methane production: a review. Biomass & Bioenergy, 13:83-114.

- Owen, W., D. Stuckey, J. Healy Jr., Young, and P. McCarty. (1979). Bioassay for monitoring biochemical methane potential and anaerobic toxicity. Water Research, 13:485-492.

- Faivor, L., and D. Kirk. (2011). Statistical verification of a biochemical methane potential test. In Proceedings of American Society of Agricultural and Biological Engineers, Louisville, Kentucky.

Anaerobic Toxicity Assays (ATA):

Anaerobic Toxicity Assays (ATAs) generally show the potential of a co-digestion feedstock to inhibit biogas production and/or methane production. Some usable ATA methods can be found in the following references:

- ISO 13641-1:2003 Water quality – Determination of inhibition of gas production of anaerobic bacteria – Part 1: General test. International Standards Organization.

- ISO 13641-2:2003 Water quality – Determination of inhibition of gas production of anaerobic bacteria – Part 2: Test for low biomass concentrations. International Standards Organization.

- Owen, W., D. Stuckey, J. Healy Jr., Young, and P. McCarty. (1979). Bioassay for monitoring biochemical methane potential and anaerobic toxicity. Water Research, 13:485–492.

Chapter 4

Feed Pre-treatment for Enhance Anaerobic Digestion

Over the recent years, options for anaerobic digestion feedstocks have increased and widen, especially with the introduction of waste to energy concept. Different wastes or feedstocks types required different pre-treatment or pre-treatment series based on their characteristics. Selection of pre-treatment plays a very significant role in the anaerobic digestion process; because not only the overall process train and process efficiency but also the final process outcomes and economics are dependent on that.

In general, the feed pre-treatment technologies are focused on achieving the following objectives:

- Removing toxic, inhibitory and unwanted substance for digester
- Reducing digester maintenance and clean up
- Homogenizing the feedstock mixture
- Adjusting moisture content, temperature etc. of the feed for digestion process
- Enhancing biogas production from the digester
- Promoting higher organic and volatile solid destruction inside the digester
- Reducing feedstocks volume and increasing in digester capacity
- Increasing hygienic safety and removing high pathogen from feedstocks
- Decreasing processing and disposal cost of the end products

Pre-treatment Categories and Related Technologies

In general, the type and level of pre-treatment mainly depends on the purity and characteristics of the feedstocks.

Commonly, the first step of any pre-treatment process is to sort out and remove the non-biodegradable and inert contaminants from the feedstocks such as grit, sand, stones, metal pieces, paper or plastic packaging etc. These contaminants, If not removed, can cause damage to the downstream processes and equipments; for example excessive grit and sand can settle in the digester bottom, clog pipes and damage process valves and sensors.

Hygienic considerations become essential when the feed steam may contain harmful microorganisms and pathogens depending on their origin and condition – such feed stokes are slaughterhouse wastes; other animal wastes like slurry from chicken, pigs and cattle farms; food wastes and wastewater sludge.

Some feedstocks, like wastewater sludge or waste rich in ligno-cellulosic component, need pre-treatment for further process enhancement and higher energy recovery. Pre-treatments expose and increase the surface area of these feedstocks by breaking them down further and thus increasing the organic loading rates and reducing the required retention time in the digester.

So, pre-treatment can be generally divided into the following three categories:

- Sorting, Separation and Homogenization
- Pasteurization and Pathogen Reduction
- Hydrolysis Enhancement and Process Improvement

Sorting, Separation and Homogenization

The first step of pre-treatment for any contaminated feedstocks is to remove non-biodegradable and inert materials that can negatively affect downstream processes and equipments.

Undesirable items in the feedstocks such as bigger piece of paper, plastic, clothing, metal pieces, rocks etc. can be manually sorted at the receiving station of the treatment plant. Then, if required, the feed stream is send to the mechanical sorters for further separation of contaminants. Feed stream like manure only require grit and sand removal while feed stream like municipal solid waste requires much more complex treatment with multiple separation units working in series. Some of the mechanical separation equipments are as follows:

- Stone trap
- Float trap
- Grit tank
- Screens
- Rotating Trommels
- Magnetic separators
- Floor Scraper
- Hydrocyclone

Some commercially available separation units (MacFarlane and Davis, 2011) are:

- Horizontal Separator (Hybag, www.hybag.ch)
- Vertical Bioseparator (Doda, www.doda.com/index.htm)
- Hammer Mill Separator (Wackerbauer, www.wackerbauer-maschinenbau.de)
- Vertical Chain Mill + Piston Press (FITEC, www.fitec.com)

Sometimes feedstocks need to be homogenized before feeding it to the digester to ensure an efficient process performance and smooth operation. Feedstocks like energy crops (grass silage etc.) are usually chopped to a uniform size. Other feedstocks may be converted to smooth slurry for feeding and mixing conveniences. Equipment involved in the homogenization phase may involve the following:

- Shredders
- Screw Cutter
- Milling
- Rotating Drums

Commercially available homogenisation systems utilize mixing and coarse grinding actions with augers and shaft based grinders. Key technology suppliers include Vogelsang (www.vogelsangusa.com/products/digester-feed-products) and Borger (www.boerger.com) and some of their chopping, grinding and feeding technologies (Black, 2012) are:

- Vogelsang Energy Jet solid material feeder
- Vogelsang Quickmix system
- Vogelsang BioCut
- Borger Multicrusher
- Borger Multichopper
- Borger Powerfeed SSR

Feedstocks homogenization sometimes also includes addition of water to reach desirable dry matter content. It is important to maintain the optimum solid content inside the digester to prevent clogging, settling or formation of floating layers.

Pasteurization and Pathogen Reduction

Any feedstock that contains pathogens or infectious microorganisms need to go through some level of pre-treatment for hygienisation, sterilization or pasteurization based on the degree of contamination.

The degree of treatments depends on both the temperature and the holding time. Sterilization can be achieved by at least 20 minutes of exposure at a core temperature of more than 133 °C and an absolute steam pressure of no less than 3 bar. In the other hand pasteurization is critical for waste containing animal by-products and required at least 60 minutes in the temperature range of   70 to 90 °C (Genesis Projects Corp., 2007). One of the commercially available hygienisation units is described below:

BioChop Hygienisation Unit (www.landiainc.com)

As per the manufacturer website, it is a completely free-standing unit for the mixing and heat treatment of the feedstocks and can be supplied in various sizes. When the feedstocks are added to the unit, it is comminuted, liquefied, heat treated and pumped back out in one process. The BioChop unit also claimed to minimise obnoxious smells and prevent pests from entering the residual products.

Hydrolysis Enhancement and Process Improvement

It is widely stated that hydrolysis is one of the rate limiting steps for anaerobic digestion especially in the presence of complex organic compounds in the feedstocks. Various pre-treatment techniques and their optimization protocols are available in the literature, although they are conducted mainly in academic and bench scale settings. Some examples are thermal pre-treatment (Li and Noike, 1992), addition of enzymes (knapp and howell, 1978), ozonation (Yasui and sibata, 1994), chemical solubilisation by acidification (Woodard and Wukasch, 1994) or alkaline hydrolysis (Mukherjee and Levine, 1992) and mechanical shearing (Kopp et al., 1997). Pre-treatment definitely has a potential to enhance the digestion and increase energy recovery in anaerobic digestion. In general, feedstocks pre-treatment method for hydrolysis and process improvement can be categorized as following:

- Physical and Mechanical Pre-treatment: chopping, grinding, milling, ultrasound, etc.
- Physico-chemical and Chemical Pre-treatment: Using chemicals such as alcohols, alkaline, acids, ozone, etc.
- Thermal Pre-treatment: Using heat and hot water.
- Biological and Enzymatic Pre-treatment: Using specialized microorganisms and specific enzyme treatment to the feed.

In the following sections, these pre-treatment categories are described briefly with one or two example from the numerous literatures available on them:

Physical/ Mechanical Pre-treatment:

Physical/Mechanical pre-treatments are mainly consisting of chopping, grinding, milling, ultrasound, microwave irradiation etc. (Taherzadeh et al, 2008; Shahriari et al, 2012) to reduce the overall particle size of the feedstocks. Mechanical pre-treatment of feedstocks provides advantages like higher exposed surface area of feed resulting in higher microbial access and degradation; increased cell rapture and release of organics in the solution resulting in better biodegradability and higher gas production.

Physico-chemical and Chemical Pre-treatment

Physico-chemical or chemical pre-treatment consist of steam explosion, application of alkaline solutions such as NaOH, $Ca(OH)_2$ (lime) or ammonia, treatment with ozone, treatment with acid such as dilute sulphuric acid, acetic acid and nitric acid etc (Taherzadeh et al, 2008). The aim of the chemical pre-treatment is to break down the complex organic compounds to simpler structures that are more hydrolysable during the digestion process.

Thermal Pre-treatment

In thermal pre-treatment feedstocks are subjected to a higher temperature in the range of 40–150 °C (Müller, 2000) to accelerate the hydrolysis process by breaking down the long chain bio-molecules. Lie et al (2012) suggested that thermal pre-treatment at 175 °C for 60 min significantly decreases viscosity, improves the dewatering performance, as well as increases soluble chemical oxygen demand, soluble sugar, soluble protein, and especially organic compounds for municipal biomass wastes. In addition, by thermal pre-treatment 59.7%, 58.5%

and 25.2% of the organic compounds can be separated in the liquid phase from kitchen waste, vegetable/fruit residue and waste activated sludge respectively. Moreover, waste activated sludge achieves a 34.8% methane potential increase and a doubled methane production rate after thermal pre-treatment.

Biological and Enzymatic Pre-treatment

Specialized microorganism or specific enzymes have also been used to enhance hydrolysis of the feedstocks. Enzymes such as lipases, proteinases, cellulases and hydrolases have been reported to to accelerate hydrolysis in the literature (Dohányos et al., 2004; Mshandete et al., 2005; Schieder et al., 2000). Several fungi, such as brown, white and soft-rot fungi, have been used for pre-treatment of lignocellulosic feedstocks and white-rot fungi are reported to be the most effective microorganisms for lignocelluloses (Sun et al, 2002). The main advantages of biological pre-treatments are lower energy requirement, no chemical requirement and mild environmental conditions; but for most biological pre-treatment processes the treatment rate is very low (Taherzadeh et al, 2008).

Some of the commercially available proprietary pre-treatment technologies are as discussed in the following section:

Micro sludge (http://microsludge.com/products/microsludge):

As per the manufacturer, Microsludge is an industrial scale homogenizer. Microsludge process destroys and liquefies the cell membrane of the feedstock and ultimately homogenizes it by utilizing a large, abrupt pressure drop (12,000 psig/82,700 kPag) and a high velocity of flow. This process thus boosts anaerobic digestion by increasing its capacity and efficiency while reducing its operating cost. The process is most suitable for waste activated sludge pre-treatment.

GBU mbH Homogenization, Suspension and Separation Unit

(http://www.gbunet.de/frame-set/Frame-Set%20BG-AUF-e.html)

GBU offers a pre-treatment unit that perform homogenization, suspension and separation in a single process. First, pre-crushed material is fed into a steel tank and diluted to approximately 10% solids. Then it is intensively mixed and suspended to achieve greater solubility of fats and proteins. For hygienisation purposes, the suspension is then heated up to 70° C. The floatable contaminants like plastics, paper etc. are removed from the top by a specially designed screen. Contaminants like glass and metals are removed from the bottom of the tank once settled. As a result a homogenous liquid of organic matter, free of contamination gets created and that can be directly pumped to the digester.

BioRefinex Process (http://www.biorefinex.com/technology.php)

The patented BioRefinex process consists of a "thermal hydrolysis reactor", which uses high pressure and saturated steam to denature organic material and destroy pathogens. The end product is a sterilized organic feedstock for the digester. As per the manufacturer's claim, this process can complete "the hydrolysis step" of the feedstock even before it enters the digester. As a result, feedstock will have shorter retention time and higher destruction rates inside the digester.

Cambi Thermal Hydrolysis Process (www.cambi.no)

Cambi's Thermal Hydrolysis Process (THP) is a high-pressure steam pre-treatment process that claims to be a suitable pre-treatment process for municipal and industrial sludge as well as for bio-waste. Under high pressure and temperature, organic compounds gets dissolved, hydrolyzed and sterilized. The highlights of Cambi process are listed as follows:

- Sludge is dewatered (16-17% dry solids) and stored in storage silo(s)
- The dewatered sludge is then send to the pulper to be mixed and heated by steam (temperature is around 97⁰C and retention time about 1.5 hour).
- The next step is thermal hydrolysis that takes place in reactor(s) at 165°C for 20 to 30 minutes.
- The sterilized sludge is then passed rapidly into the flash tank causing cell destruction from the pressure drop.
- Before sending to the digester, the sludge is cooled to the required digestion temperature partly by adding dilution water and partly by using heat exchangers.

Some of the benefits claimed by the Cambi process include:

- Robust anaerobic digestion process
- Enhanced biogas production
- Improved dewaterability after digestion by 50% - 100%
- Pasteurization and stabilization of final biosolids product/cake
- The digested sludge has no negative odour

BTA® Hydromechanical Pre-Treatment (http://www.bta-international.de/en/home.html)

The BTA® Hydromechanical Pre-treatment facilitates claims to remove impurities as well as transfer digestible organic components into an organic suspension suitable for anaerobic digestion. The process has two core components, the BTA® Waste Pulper and the BTA® Grit Removal System.

The BTA® Waste Pulper

In BTA® Waste Pulper, the feedstock is separated into fractions by utilizing natural buoyancy and sedimentation forces of its components. First, feedstock is added to the BTA® Waste Pulper pre-filled with process water. The heavy fraction settles into the bottom and the light fraction floats up. In addition, the non-soluble organic components are also reduced to fibers by shearing forces and then brought into suspension. At the end, settled heavy fractions are collected from the bottom and floated light fractions are skimmed off from the top. As a result, three fractions are effectively produced and separated from this process:

- Digestible organic materials
- Light fraction: plastics, foil, textile, wood, etc.
- Heavy fraction: stone, bones, batteries, etc.

The BTA® Grit Removal System

The processed organic fraction of BTA® Waste Pulper may still contain some sand and fine impurities, that can be safely removed by the BTA® Grit Removal System. The efficient sand removal by this process will provide further protection of downstream plant components from wearing, silting up, obstruction and sedimentation.

Pre-treatment of Food Waste and Organic Fraction of Municipal Solid Waste

Effective pre-treatment of food waste and organic fraction of municipal solid waste (OFMSW) are essential to ensure efficient performance of the digestion process. These wastes need to be free of contamination like plastic and other packaging materials, cardboard, metal pieces, wood particles, glass pieces etc., and then they need to be converted into a homogeneous organic mixture suitable for anaerobic digestion process. Depending on the level of impurity and sources of the products, food waste/OFMSW feedstock may be required the following pre-treatment technologies for the purification:

- Receiving and manual sorting
- Bag opener
- Shredders
- Screw Cutter
- Milling
- Rotating Drums
- Rotating Trommels
- Screens
- Magnetic separators
- Sterilizer
- Pulper
- Hydrolyser
- Homogenizer

Example of Food Waste/ OFMSW Pre-treatment Plants and Technology Suppliers

BioPrePlant® System (www.biopreplant.com)

The BioPrePlant® System is designed for pre-treatment of food waste. BioPrePlant System also has a customized automation and control system to reduce operator interventions. The BioPrePlant® System has the following stages and components (http://www.biopreplant.com/Resources/BioPrePlant-System_Folder_EN.pdf):

Reception

BioPrePlant-System's food waste reception is a multi-waste system where food waste from households, supermarkets, restaurants and other wastes are collected together without any sorting or de-packaging. The mixed food waste is then automatically transported for further treatment and separation.

BioPreCrusher

In the next stage of the process, packaging bags and other food containers such as tins are opened and ferrous metals are removed by using the BioPreCrusher machine. The BioPreCrusher is very important part of the BioPrePlant-System as it prevents undesirable metals and other larger objects from entering into the process. The BioPreCrusher contains a crusher, a transport conveyor and a magnet. In this step, larger non-crushable objects are automatically discharged from the machine as well.

Material Transportation

Material transportation in a BioPrePlant-System is conducted via spiral conveyors or pumps to ensure a clean and efficient system. As per manufacturer claims, it also minimizes spillage and leakages during operation that improves the overall working environment and surroundings.

BioSep® Stage 1

The next stage is a patented process machine called BioSep. It is dedicated to separate plastics and packaging material from mixed food waste. This machine separates, washes, and dries plastics and food packaging material and discharges them out of the machine as reject.

BioSep® Stage 2

BioSep Stage 2 has the similar function as BioSep Stage 1. The difference is that BioSep Stage 2 has a finer masking of the sieve that ensures separation of even small pieces of plastics and packaging material that wasn't separated in BioSep Stage 1 in order to utilize the resource in food waste optimally.

Biosubstrate

A biosubstrate suitable for anaerobic digestion is the end product from a BioPrePlant-System. Manufacturer claims that the particle sizes of the organic material, and the amount and sizes of the remaining impurities are well within the technical and environmental requirements for biogas production.

Pre-treatment Plant by SpiralTrans (http://www.spiraltrans.com/en/)

SpiralTrans uses a number of proprietary process components to produce a biosubstrate feed for digester that is free from plastic and other unwanted reject fractions. They claim that both the biosubstrate and reject fractions are of high quality that can comfortably meet the demands.

The pre-treatment plants receive food waste in receiving bunkers. Then they get transferred by spiral conveyors to the BagTronic® plant, which is an automatic sorting system that separates waste bags of different colours for further processing and handling. Then they get sieved, purified, grinded and mixed in the next several stages by different spiral screens, grinders and magnetic separators etc. SpiralTrans uses OrgaSep© process for further purifying the digester feed. The OrgaSep© separates plastic and other contaminants from the biosubstrate to produce an end product that is a purified and pumpable.

Sysav's Pre-treatment Plant for Food Waste (http://www.sysav.se/In-English1/The-Sysav-concept/)

In Sysav Biotec's pre-treatment plant, food waste is converted into pumpable, viscous slurry as the end product. Sysav's pre-treatment plant can receive three type of food waste and they receive pre-treatments accordingly:

Pumpable Liquid Food Waste

Pumpable liquid food waste can be pumped directly from the transporting tanker to a receiving tank in the plant. This waste can be collected from disposal grinders and fat separators of restaurants and large scale kitchens. There is an intermediate tank between the tanker and reception tank for stones and grit removal purpose.

Pre-packed Liquid Food Waste

Pre-packed liquid food waste inside plastic or paper packaging, such as juice and milk in cartons, can come from food producers and wholesalers. The packaging of this waste has to be punctured to collect the liquid inside and this is done via a piston press. The squeezed liquid is collected in a collection tank and the compressed packaging is then sent to a waste bunker via screw conveyor.

Separated Solid Food Waste

Source separated solid food waste can come from households, restaurants and large-scale kitchens. The organic part of this waste need to be collected and this is carried out using a screw extruder. First the waste is transported from the receiving bin via a screw conveyor and ladder conveyor to a shredder. Once in the shredder, the food wastes get shredded into smaller pieces. Then the shredded food waste is homogenised in a mixing unit. Sometimes the food wastes need to be diluted and this dilution is done by adding other liquid food waste or process water. In the screw extruder, the waste is pressed onto a cone that provides counter-pressure. As a result thick organic slurry extruded through the sides of the machine. This slurry is then use for biogas production.

Chapter 5

Process Consideration, Configuration and Technology Selection for Enhanced Anaerobic Digestion

Process Considerations for Optimal Performance

There are several important growth kinetics, environmental factors and operating parameters that have to be controlled in order to optimize anaerobic digestion process. These parameters along with their optimum ranges, as reported in various literatures (Metcalf and Eddy, 2003; Olvera and Lopez, 2012; Seadi et al, 2008; Cavinato, 2011) are briefly discussed below:

- Temperature
- Solids and Hydraulic Retention Time
- Alkalinity
- pH
- Toxic Substances
- Carbon and Nutrients Availability
- Organic Loading Rate (OLR)
- Product Concentrations

Temperature

In anaerobic digestion process, temperature is not only important for microbial metabolic activities but also for the overall digestion rate, specifically the rates of hydrolysis and methane formation. In general, anaerobic digestion process can occur within a wide range of temperatures. This temperature range has been divided into three groups: psychrophilic– less than 20 ^{0}C, mesophilic– 30 to 42 ^{0}C and thermophilic– 43 to 55 ^{0}C (Seadi et al, 2008). In practice, most of the anaerobic digestion system are designed to operate at mesophilic range, between 30 to 38 ^{0}C, and some of them are designed for thermophilic temperature range of 50 to 57 ^{0}C (Metcalf and Eddy, 2003). In general, thermophilic digestion processes potentially allow higher loadings with reduced hydraulic retention times, higher conversion efficiencies and pathogen disinfection while mesophilic digestion is more stable, less at risk from ammonia nitrogen toxicity and requires less process heat (Yirong et al., 2013).

Many modern anaerobic digestion plants have chosen thermophilic process temperatures as the thermophilic process provides some advantages, compared to mesophilic and psychrophilic processes, although they have some disadvantages as well (Seadi et al, 2008):

Advantages of thermophilic process

- Destroy pathogens effectively
- With reduced retention time, the digestion process become faster and more efficient
- Provide better degradation and utilization of substrates
- better potential for solid liquid separation

Disadvantages of thermophilic process:

- Higher degree of imbalance
- Higher energy demand due to high temperature
- Higher risk of ammonia inhibition

Solids and Hydraulic Retention Times

Solids and Hydraulic Retention Times (SRT and HRT) are the average time solids and liquids are held in the digestion process. Anaerobic reactions (hydrolysis, fermentation and methanogenesis) and anaerobic reactor size are directly related to these parameters. Each of the anaerobic digestion reactions requires a minimum SRT to be completed and if the design SRT is less, than that the digestion process will fail (Metcalf and Eddy, 2003). In a completely mixed reactor with no recycle, solids and hydraulic retention times are the same. In practice, for high rate digestion the values of SRT range between 10 to 20 days (Metcalf and Eddy, 2003).

Alkalinity and pH

In general, hydroxides and carbonates of calcium, magnesium, sodium, potassium and ammonium produce alkalinity in the wastewater. Alkalinity plays an important role in anaerobic digestion process as it control the pH by buffering the acidity created in the acidogenesis process (Olvera and Lopez, 2012). The alkalinity of the digester in general is proportional to the solids feed concentration of the digester and in a well established digester the total alkalinity ranges between 2000 and 5000 mg/L (Metcalf and Eddy, 2003).

The growth of anaerobic process microorganism significantly depends on the pH value of the system. Most methanogens prefer a narrow pH range and the optimal is reported to be 7 to 8.

Acidogens usually have a lower value of optimum pH. The optimum pH interval for mesophilic digestion is between 6.5 and 8 and the process is severely inhibited if the pH value falls out of this range (Seadi et al, 2008).

Toxic Substances

Presence of toxic inhibitory compound can adversely affect the anaerobic process microorganism. A wide variety of inorganic and organic toxic and inhibitory substances can cause anaerobic digester upset or failure. The commonly present toxic substance in anaerobic digesters include ammonia, sulfide, light metal ions, heavy metals etc. (Chen and Creamer, 2008).

Some inhibitory substance as explained in 'Digestion Inhibitors' documents by The Sustainable Energy Authority of Ireland (SEAI–http://www.seai.ie/) are listed as follows:

- Oxygen and light; high amount of them could inhibit the activity of methane producing bacteria.
- Disinfectants such as herbicides, heavy metals or antibiotics found in poultry/chicken manure can also disturb the process if present in high concentration.
- Hydrogen sulphide (H_2S) is a product of the digestion process but can be found in the organic material as well. Hydrogen sulphide and sulphuric acid are highly corrosive and could seriously affect the components of digester.
- A high ammonia concentration, which is a cellular poison, could be caused by high nitrogen and ammonium (NH4) concentrations under certain circumstances. In such case, substrates with a high Nnitrogen concentration like chicken manure/pig slurry should be diluted or mixed with another nitrogen–poor substrates.

Carbon and Nutrients Availability

Nutrients like carbon, nitrogen, phosphorus and sulphur are very important for the survival and growth of anaerobic digestion process organism. Different micronutrients/microelements (trace elements) like iron, nickel, cobalt, selenium, molybdenum or tungsten are also essential for the anaerobic process microorganisms. Insufficient amount of these nutrients and trace elements can cause inhibition and instability in anaerobic digestion process. The ideal carbon to nitrogen (C: N) ratio for anaerobic digestion ranges from approximately 20:1 to 30:1 (EPA, 2012). The optimal nutrient ratio for the carbon, nitrogen, phosphor, and sulphur (C: N: P: S) is considered to be 600:15:5:1 (Seadi et al, 2008). It is also reported that to maintain optimum methanogenic activity, desirable liquid phase concentration of nitrogen, phosphorus and sulphur should be in the order of 50, 10 and 5 mg/l (Metcalf and Eddy, 2003). In addition, it is suggested that level for iron, cobalt, nickel and zinc should be 0.02, 0.004, 0.003 and 0.02 mg/g acetate produced respectively (Metcalf and Eddy, 2003).

Organic Loading Rate (OLR)

Organic Loading Rate (OLR) is defined as the amount of organic dry matter that can be fed into the digester per unit volume of its capacity per day. It is usually calculated based on the mass of volatile solids added per day per unit volume of digester capacity. Another way of calculating it is, the amount of volatile solids added to the digester each day per mass of volatile solids in the digester; although the first approach is favorable (Metcalf and Eddy, 2003). Loading rate is an important operational factor for digester because if it is too high, valuable methane former can washout from the system. In addition, toxic materials like ammonia can accumulate and upset the process. On the other hand, if the lading rate is too low, it can result in lower organic solids destruction and lower biogas production. Moreover, larger uneconomical digester will require higher heats. For these reason, the optimum loading rate should be a compromise between the highest possible biogas generation and having a justifiable plant economy (Seadi et al, 2008).

Product Concentrations

The stability of the anaerobic digestion process also depends on concentration of some products produced during the organic break-down process, like Volatile Fatty Acid (VFA). During the acidogenesis process, different fatty acids like acetate, propionate, butyrate, lactate etc. are produced. Excessive accumulation of these acids can drop the pH value inside the reactor when the buffering capacity of the digester is exhausted. The buffering capacity of the digesters and how they will react to certain amount of VFA concentration vary from digester to digester based on its microbial population (Seadi et al, 2008).

Process Selection and Configurations for Enhance Performance

There are many different anaerobic digestion systems available in the markets that vary in process configurations and operating conditions. The design considerations and operating conditions of these treatment trains may be suitable for a particular type of feedstock or feedstocks mix but may not be applicable or economical for others. That is why digester type and technology should be selected based on the indented feedstock's characteristics and availability (the amount to be treated) as well as desired output and process economy in mind.

A brief discussion on these different anaerobic digestion technologies based on their feed type, operating conditions and process configuration are discussed as follows:

Process Selection Consideration - Feedstock's Solid Content

New and emerging feedstocks like organic fraction of municipal solid waste (OFMSW), energy crops, food waste etc., have been introduced to the anaerobic digestion technology. The concept of co-digestion or digesting two or three feedstocks types together is also getting popular. As a result the type and solid content of the feedstock mix is getting crucial for selecting the anaerobic digestion process and technologies. Based on the ability to handle the feedstocks solid content, anaerobic digestion technology can be broadly divided in to two categories–

1. Low Solid/ Liquid Phase/Wet Digestion Technology
2. High solid/Solid State/ Dry Digestion Technology

Low Solid/ Liquid Phase/Wet Digestion

Anaerobic digestion systems that can handle input solid content only up to 15% are known as low solid anaerobic digestion. These are the most common type of anaerobic digesters available. They are also called wet digestion or liquid–phase digestion. The types of feedstocks treated by this group of digesters are wastewater sludge, industrial wastewater, manure, food waste slurries, FOG etc.

High solid/solid State/ Dry Digestion

Anaerobic digestion systems that can handle higher solid content (>15%) are known as high solid digestion or dry digestion. Dry digestions are suitable to handle waste at high solid loadings as well as lingnocellulosic biomass such as silages of corn, grass, rye etc.; coarsely shredded MSW, residential food waste along with lawn and yard waste etc. This type of digesters usually produce end product with higher solid content and consume less energy for heating. But they require longer retention time and additional equipments for mixing and material flow on the output side. The high solids anaerobic digestion that is designed for pumpable high solid feed like wet food waste slurries (usually 15–35% solid content) are further known as wet high solid or dry pumpable digestion.

Some dry digestion technologies are briefly described below:

DRANCO Anaerobic Digestion Technology (http://www.ows.be/household_waste/dranco/)

The DRANCO anaerobic digestion technology by Organic Waste Systems (OWS) is a patented process. The three main characteristics of DRANCO are its vertical design, high–solids concentration and the absence of mixing inside the digester. DRANCO digestion is often applied to mix waste or residual household waste or organic fraction of municipal solid waste (OFMSW), source separated organics, yard and food waste. The plant can operate at dry matter content inside the digester of up to 40 %. Some characteristics of this technology as reported by OWS are:

- High rate dry digestion
- Vertical digester with feeding at the top and removal from the bottom
- Single phase digestion with recycling
- Option for thermophilic or mesophilic operation
- No mixing, stirring or gas injection in the digester
- Single digester with a capacity up to 60,000 tons per year of household organic waste.

Kompogas Anaerobic Digestion (http://www.axpo.com/axpo/kompogas/en/home.html)

The Kompogas is a patented dry anaerobic digestion process for organic waste by Axpo Kompogas Engineering Ltd. The process uses a continuously feed horizontal plug–flow digester. The feed system automatically conveyed the feed to the digester and process water is added inside the digester to maintain the optimum ratio. The digester has a low speed agitator inside for mixing and preventing sedimentation. The temperature of the process is thermophilic operating at 53–55 $^{\circ}$C. The average feedstock moisture content is around 75% and the retention time inside the digester is approximately 14 days.

GICON High Solid Anaerobic Digestion (http://www.gicon-engineering.com/en/about-us.html)

GICON is a two-stage dry-wet anaerobic digestion process and operates in batch mode. The technology was developed, tested and commercialized in Germany by Grossmann Ingenieur Consult GmbH (GICON). The Harvest Energy Garden located in Richmond, BC, Canada is based on this technology and is owned and operated by Harvest Power Canada Ltd. This is the first industrial-scale high-solids anaerobic digestion (HSAD) plant in Canada and is also one of the largest HSAD plants in North America. The plant processes approx. 30,000 t/a of combined food waste and yard waste and produces approximately 770 kW of electrical energy (www.gicon.de/uploads/tx_sbdownloader/Factsheet_Vancouver_EN.pdf).

Eisenmann High Solids Anaerobic Digestion System (http://www.eisenmann.us.com/biogas-renewable-energy-waste-to-energy/)

Eisenmann has a modular high solids anaerobic digestion system. The system is typically based on a two-stage process with a primary and secondary digester. The main characteristics of this process are:

- Continuously mixed plug-flow process
- Enclosed system with potential odor prevention
- Modular design

The BIOFerm™ Dry Fermentation System (http://www.biofermenergy.com/bioferm-system/)

The BIOFerm™ is a batch type dry anaerobic digestion system that operates in the mesophilic temperature range. It is well suited for solid waste materials with a solids content of 25-35%.

BIOFerm plants consist of multiple concrete garage style digesters that can be sealed gas-tight. Front-end loaders load feedstocks into these chambers where they get digested for an average of 28 days. The digesters are heated through in-wall radiant heat and through the percolate, which is sprayed onto the feedstocks using the over-head sprinklers. Inside the digester, feedstocks remain still. But the percolate and water produced are continuously collected through floor drains and re-circulated. The BIOFerm™ Dry Fermentation system is suitable for municipal organic waste including yard waste.

Process Selection Consideration - Digester's Operating Temperature

Mesophilic Anaerobic Digestion

The anaerobic digestion technology that operates at the mesophilic temperature range (35–37 °C) is known as mesophilic anaerobic digestion. Mesophilic anaerobic digestion is most common system which has a more stable operation but a lower biogas production rate. Another disadvantage of mesophilic digestion is that it does not reduce the pathogen concentrations enough to produce Class A biosolids, a biosolids that contains no detectible levels of pathogens (WEF, 2004).

Thermophilic Anaerobic Digestion

The anaerobic digester that operates at the higher thermophilic temperature range (50–65°C) is known as thermophilic anaerobic digestion. Interest in the thermophilic digestion developed based on the facts that higher temperatures reduce pathogens and thermophilic temperatures provide more rapid reaction rates than mesophilic temperature. Class A quality biosolids can be produced from thermophilic digestion when the time temperature criteria specified in the US EPA Part 503 are satisfied. Thermophilic anaerobic digestion in general are more efficient in biogas production but associated with higher maintenance cost.

The advantages of thermophilic digestion listed in the WEF white paper on the anaerobic digestion (2004) are:

- Increased volatile solids reduction
- Faster reaction rates for shorter retention times
- Higher capacity for a given volume
- Increased pathogen destruction
- Improved dewaterability of the digested biosolids

The most frequently cited disadvantages of thermophilic digestion are:

- Higher odor formation resulting from a higher volatile fatty acid (VFA) concentration
- Higher energy requirements for heating
- Increased sensitivity to thermal shock

Process Selection Consideration - Digester's Stages and Rector Configuration

Single Stage Anaerobic Digestion

In a single stage anaerobic digestion, all the anaerobic process reactions take place inside a single reactor. The operating conditions are more or less suitable for all the reactions and no particular phase has been optimized. Single stage anaerobic digestion requires less capital cost, less maintenance but has lower gas production and organic conversion rate.

Multi- Stage Anaerobic Digestion

In multi-stage anaerobic digestion multiple reactors (usually two of them) are designed in series to optimize the process and enhance gas production. By introducing physical separation or staged operation, facilities can optimize each process reaction for the breakdown of the organics and enhance more organic destruction. Although multi-stage digestion process is more complex and require more capital cost, some of the benefits of this system as reported (EPA, 2006) are as follows:

1. Enhance gas production
2. More volume reduction through VS destruction
3. Higher quality biosolid (Class A biosolid can be achieved by incorporating a thermophilic stage)
4. Better odour control
5. Can configure to reduce foaming problem
6. Short circuit prevention via multiple reactors, each with optimized retention time.

Some configurations of the multi-stage anaerobic digestion (EPA, 2006; Metcalf and Eddy, 2003) are listed and discussed below:

*   Staged Mesophilic Digestion
*   Acid/Gas (AG) Phased Digestion
*   Temperature Phased Anaerobic Digestion (TPAD)
*   Staged Thermophilic Digestion

Staged Mesophilic Digestion

Two anaerobic digesters coupled in series as primary and secondary digester is not uncommon, but there is not much information available regarding two heated well mixed digester in series. Literature findings indicated that two staged mesophilic digestion may produce more stable, less odorous biosolids that are easier for dewatering (Metcalf and Eddy, 2003).

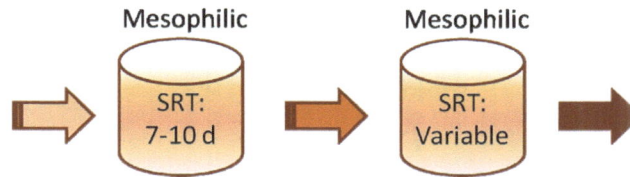

Figure 5.1: A Staged Mesophilic Digestion Configuration
(Data Source: Metcalf and Eddy, 2003)

Acid/Gas (AG) Phased Digestion

In a two-stage acid/gas (AG)-phased digestion, the acid-forming stage (hydrolysis and volatile acid fermentation) are physically separated from the methane gas-forming stage by dedicating a digester designed and optimized for each of these two stages. The first digester is commonly known as primary or acid phase digester. The first phase is conducted at an acidic pH (pH 6 or less) and at a lower SRT favorable for volatile acid formation. In the second stage digester, usually known as secondary digester, methanogenic bacteria convert the acids and soluble organics into methane rich biogas. The second digester is operated at a neutral pH and longer SRT to suit the need of methanogenic bacteria and maximize gas production (Metcalf and Eddy, 2003). In a typical two stage AG system, primary digester is usually heated to optimize the hydrolysis and acid forming reaction, while the secondary digester is not because of the exothermic (heat-producing) nature of the methane formation reaction (EPA, 2006). Some advantages and disadvantages of the AG process, as listed in the WEF white paper on the anaerobic digestion (2004) are:

Advantages:

- Greater volatile solids destruction in a smaller volume than a conventional mesophilic digestion process

- A reduction in digested solids processing and handling
- Foam control
- Increased gas production and stability
- Class A potential with a thermophilic reactor operation (AG–MT/AG–TM)

Disadvantages:

- Increased complexity from multiple phases

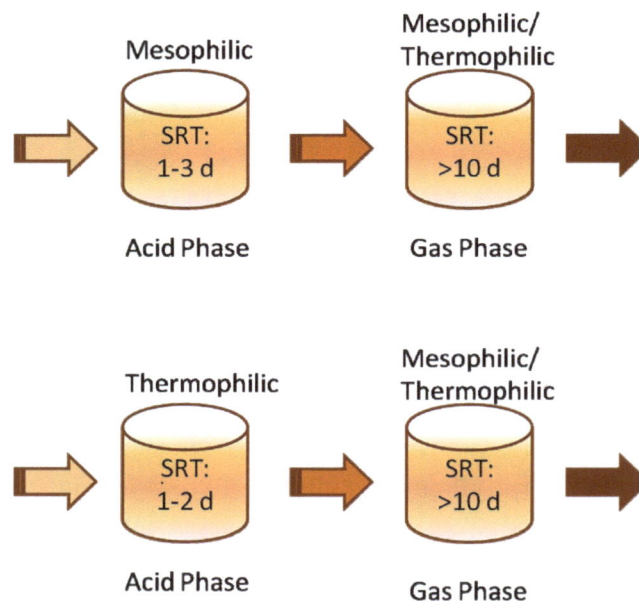

Figure 5.2: Two Acid/Gas (AG) Phased Digestion Configurations
(Data Source: Metcalf and Eddy, 2003)

Temperature Phased Anaerobic Digestion (TPAD)

The Temperature Phased Anaerobic Digestion (TPAD) incorporates both thermophilic and mesophilic reactors connected in series to take advantage of the higher digestion rate of thermophilic digestion as well as the higher stability of the mesophilic digestion. The thermophilic digestion rate is usually four times faster than the mesophilic digestion (Metcalf and Eddy, 2003). On the other hand, adding a mesophilic digestion at the end not only improved stability of the digestion operation but also destroyed odorous compounds produced during the thermophilic stage. Overall the TPAD process has greater volatile suspended solids (VSS) destruction and gas production ability. It is reported that the VSS destruction efficiencies of the TPAD process are 15–25% greater than the single stage mesophilic digestion process. The TPAD process is also reported to have better shock absorbing capacity, lesser foaming issues and the ability to produce a 'Class A' biosolids (Metcalf and Eddy, 2003).

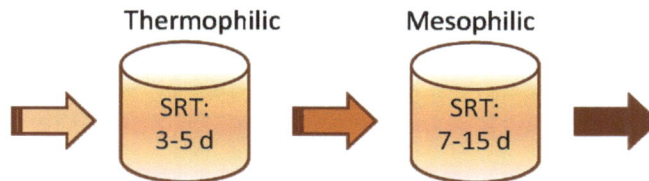

Figure 5.3: A Temperature Phased Anaerobic Digestion (TPAD) Configuration
(Data Source: Metcalf and Eddy, 2003)

Staged Thermophilic Digestion

In staged thermophilic digestion, a large digester is followed by one or more smaller digester to prevent pathogen short circuiting and achieve a Class A biosolid. One such example is Annacis Island Wastewater Treatment Plant in Vancouver BC, Canada; where first stage is followed by three subsequent stages (Metcalf and Eddy, 2003).

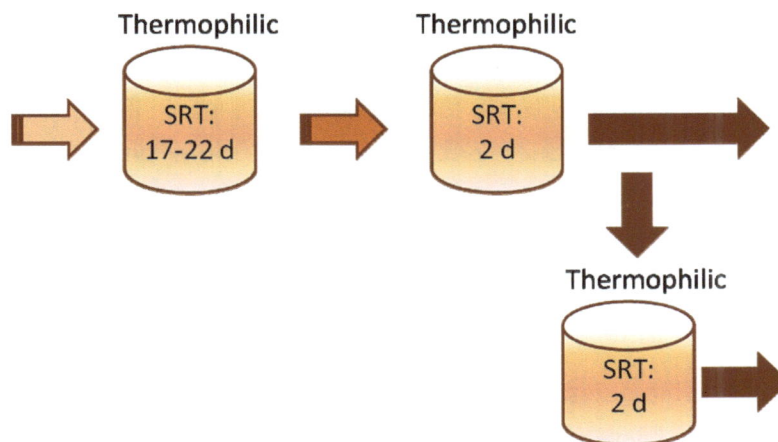

Figure 5.4: A Staged Thermophilic Digestion Configuration
(Data Source: Metcalf and Eddy, 2003)

Process Selection Consideration - Digester's Feeding Mode

Continuous Feed Digester

Continuous feed digesters are either feed continuously or semi-continuously. These are the most common type of digesters.

Batch Feed Digester

In Batch feed digesters, feedstocks are loaded in the digester and left there for a certain period for digestion to take place. Usually the batch digesters need to be bigger in volume due to the long retention time.

Process Selection Consideration – Enhancing Solids Loading and Digestate Recirculation

Extended Solids Retention (ESR) Digestion

The Extended Solids Retention (ESR) digestion process has greater solids retention time (SRT) than the Hydraulic Retention Time (HRT). This is done by thickening the digestate via thickening equipment and then recycling it back to the digester. This process allows for a longer SRT than the HRT as the solids have been separated from the water and reintroduced to the digester. This extended SRT increases solid reduction and generates more biogas. Some advantages and disadvantages of this process, as listed in the WEF white paper on the anaerobic digestion (2004) are:

Advantages:

- Decreased Anaerobic digester volume
- Extended organic conversion to methane
- Potential for reduction in overall polymer usage because of recycling
- Reduced digester volume requirements may result in lower life cycle cost
- Separating the SRT from the HRT provides greater flexibility in solids removal
- Increased solids concentration in the dewatering feed may result in improved dewatering

Disadvantages:

- Additional thickening equipments
- Added complexity
- The added capital and operating cost for the additional thickening equipment.

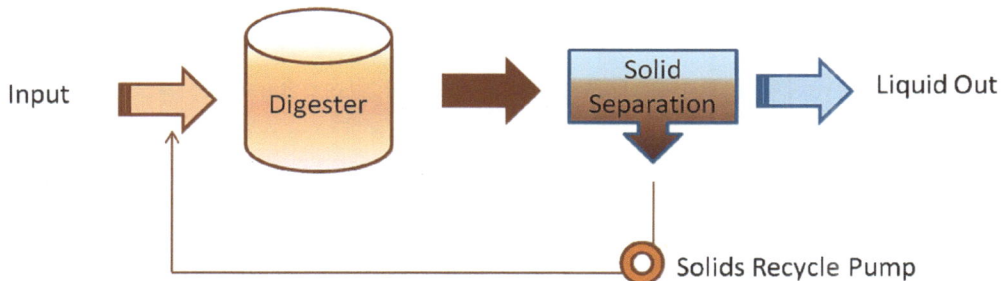

Figure 5.5: An Extended Solids Retention (ESR) Digestion Configuration
(Data Source: WEF, 2004)

Table 5.1: Example of Facilities with Multi-Stage Anaerobic Digesters

Plant	System Type
Woodridge WWTP, DuPage County, IL	Two stage AG–MT
Elmhurst, IL	Two stage AG–MM
Back River, Baltimore, MD (pilot)	Two stage AG–MM
Inland Empire (RP–1), Ontario, CA (farm manure)	Three–stage AG–MTM
Waterloo, IA	Two–stage TPAD–TM
Waupun, WI	Two–stage TPAD–TM
Rockaway, NY	Two–stage TPAD–MT
Pine Creek WWTP, Calgary, Alberta, Canada (pilot)	Three stage TPAD (multiple options being researched)
Tacoma, WA Heated aerobic stage (71∘ C [160∘ F]) +	Heated aerobic stage (71∘ C [160∘ F]) + Three–stage TPAD–TMM

M – Mesophilic stage, T– Thermophilic stage
Data Source: EPA (2006), Biosolids Technology Fact Sheet–Multi–Stage Anaerobic Digestion

Different Digester Types with Some of Their Operating Characteristics

Some of the commonly operating digester types along with their few characteristics as classified and described in EPA, 2011 document– 'Recovering Value from Waste–Anaerobic Digester System Basics' are summarized in the following Table; however, these operational conditions widely vary due to regional and site–specific considerations.

Table 5.2: Common Digester Types with Some of Their Operating Characteristics

Digester Types	Description	Percent solids	HRT	Co digestion
Plug Flow Digester	The digester tanks are usually long and narrow, typically heated and constructed below ground. Contents move through the digester as new feed is added. Some of the modified plug-flow systems can use vertical mixing techniques. These systems work best with dairy manure.	11 to 13%	15+ days	not optimal
Complete Mix Digester	Digester tanks can be either above or below ground and can be heated or unheated with impermeable gas collecting cover. The digester content gets mixed by mixers or pumps. The complete mix digesters work best when there is some dilution of the feed.	3 to 10%	15+ days	yes
Covered Lagoon	Lagoon can be either In-ground earthen or lined, with impermeable gas-collecting cover. Contents can be heated or mixed. Covered lagoons work best with manure handled via flush or pit recharge collection	0.5 to 3%	40 to 60 days	not optimal

Digester Types	Description	Percent solids	HRT	Co digestion
	systems.			
Up–flow Anaerobic Sludge Blanket (UASB)/ Induced Blanket Reactor (IBR)	Digesters are high–rate, above-ground and heated vertical tanks. Usually the feed is added continuously to the bottom of the reactor. Bacteria are suspended in the reactor due to the flow of the feed. These systems works best for consistent, homogenous waste streams.	UASB <3%, for IBR 6 to 12%	Typically 5 days or less	yes
Fixed Film/ Attached Media Digester/ Anaerobic Filters	These are above ground, heated tank and contain media such as plastic or wood chips on which bacteria attach and grow. Feeds pass through the media and get digested as it comes into contact with the bacteria attached to the media. These digesters work best with manure in temperate and warm climates.	1 to 5%	Typically 5 days or less	yes
Anaerobic Sequencing Batch Reactors (ASBR)	These digesters are typically an above ground, heated tank with an impermeable roof that collects gas. Feed is added and removed in batches. There are four phases in the ASBR cycle: fill, react, settle, and decant. An ASBR is best suited for treating dilute wastes (i.e., manure handled via slurry).	2.5 to 8%	Typically 5 days or less	yes
High–Solids Fermentation	High solids digesters are usually heated, above ground and airtight container, designed for high solids manure and other organic substrates	18%+	2 to 3 days	yes

Digester Types	Description	Percent solids	HRT	Co digestion
	(e.g., silages such as corn, grass, or rye; food waste etc.).			

Data Source: EPA (2011), Recovering Value from Waste-Anaerobic Digester System Basics

Co-digestion Process Consideration

Feedstock Reception, Pre-Treatment and Feeding

Majority of the co-digestion feedstock are delivered via special container or truck. Unloading of the material should be done in an enclosed building for odour controlling purpose. Usually the building air is collected and treated via bio–filter. Multiple categories of feedstocks may require multiple waste storage tanks/bins etc., based on their characteristics. From the receiving, the wastes are sent to various cleaning and pre-processing units based on their requirements.

Figure 5.6: A Waste Reception Hall
Source: http://lancashire.gov.uk/environment/waste/photographs/index.asp

The most economical is to use clean or source separated organics as feeds that can go away with minimal pre-treatment. The dry high solid digestion process (TS>20%) has high contaminant tolerance and usually require less pre-treatment. But the wet low solid digestion process (TS <15%) demand higher contaminants removal due to potential damage like pipe blockage, breakage of pump or mixing devices etc. The pre-treatment of feedstocks may contain three basic steps:

1. Size reduction through chopping, sieving etc.
2. Removal of indigestible components like metal, glass, stones, paper and plastic packaging etc.
3. Hygienisation and Homogenisation.

The pre-treatment processes have been discussed in details in Chapter 4.

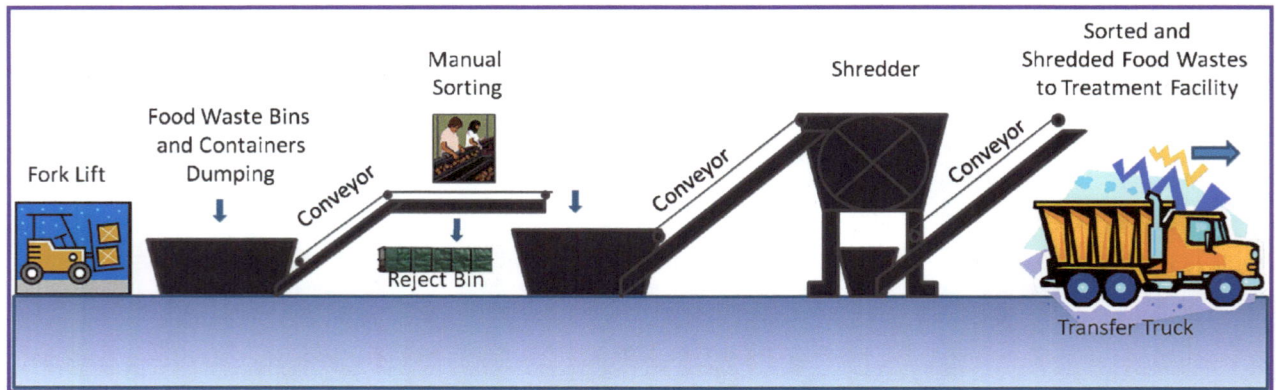

Figure 5.7: A Food Waste Separation System in a Waste Transfer Facility
(Data Source: USEPA)

Since many co-substrates cannot be fed directly into the digester, one or multiple pre-mixing and homogenizing tanks need to be provided. From this buffer tank pre-processed and pre-mixed feeds are sent to the digester based on their loading ratio.

Co-digestion Operating Conditions and Challenges

Co-digestion Feedstock Blend Ratio

Although Co-digestion is simultaneous digestion of multiple feedstocks, the most common situation is when a major amount of main basic feedstock like sewage sludge or manure is mixed and digested together with minor amounts of single or multiple high strength wastes like energy crops, food waste or FOG. A variety of such high strength wastes are is suitable for co-digestion as long as their mixtures or blending ratio with biosolids or manure are correctly optimized. The optimum blend ratio of these high strength wastes, as suggested in various literatures (shown in Table 5.3) varies from 5 to 30% of the total feed VS to the digester.

Table 5.3: Optimum Blend Ratio of Some Co-digestion Feedstocks

Co-digestion Feedstock	Blend Ratio	Digester Type	References
Grass silage, sugar beet tops and oat straw with cow manure as main feedstock.	The highest specific methane yields were obtained when feed with 30% of crop in the feedstock.	Semi-continuously fed laboratory scale CSTR Reactors.	Lehtom aki et al. (2007)
Industrial by-products of potato stillage and potato peels with pig manure as basic feedstock.	The results suggest that successful digester operation can be achieved with feed containing potato material up to 15–20% of the feed VS.	Semi-continuous co-digestion at loading rate of 2 kg VS m^{-3} day^{-1} in continuously stirred tank reactors at 35 °C.	Kaparaju and Rintala, 2004
Co-digestion of sludge from grease traps and sewage sludge.	The addition of grease trap sludge to sewage sludge digesters was seen to increase the methane yield of 9–27% when 10–30% of sludge from grease traps (on VS-basis) was added.	Both in laboratory batch and continuous pilot-scale reactor.	Davidsson et al., 2007
Food waste and wastewater biosolids	Blend ratio ranged between 5–30%.	In full-scale applications	Erdal et al., 2005

Feedstock Toxicity and Loading Rate

The feedstocks of co-digestion must not be toxic or create an environment where toxic gases are produced for digester process microorganisms. With the increased nutrients in the digester from co-digestion feedstocks, there could also be an increase in toxic gases like ammonia or hydrogen sulfide and they can inhibit the production of methane. A high organic loading rate (OLR) can cause metabolite inhibition (ammonia or hydrogen sulfide) or fatty acid accumulation which reduce the digester performance. OLR values of no higher than 0.10 1bVS/cf-d (1.6 kg/m3-d) are typically recommended to avoid such inhibition especially during the start up periods (Erdal et al., 2005).

Operating temperature

Based on the findings of Erdal et al., 2005; the number of mesophilic digesters using co-digestion substantially exceeds thermophilic digesters in full-scale operation. The main reason may be the plant economy as the additional gas production in thermophilic operation may not offer a reasonable payback to compensate the capital investment. But thermophilic anaerobic digestion may provide slightly higher methane content in digester gas than mesophilic anaerobic digestion when only food waste is fed (EPA, 2008).

pH

The necessary pH for anaerobic digestion ranges between 6.8 to 8.5 (EPA, 2012) and varies at different stages of anaerobic digestion process. Lower pH conditions inhibit biogas production because the methane bacteria cannot survive at acidic condition. Some co-digestion feedstocks like food waste tend to decompose quickly and can decrease the pH of the digester. In such case, addition of a buffer, like sodium bicarbonate may be added to balance the pH.

Nutrient balance

Anaerobic digestion process failure is sometimes linked to the imbalance of carbon and nitrogen ratio in feed mixture. In order to reduce ammonia inhibition or volatile fatty acid accumulation, C/N ratio of co-digestion feed mix should be maintained. The ideal carbon to nitrogen (C:N) ratio for anaerobic digestion is suggested approximately 20:1 to 30:1 (EPA, 2012).

The addition of co-digestion materials with higher carbon contents than the main feedstock like manure can improve the overall C:N ratio and increase methane production. For example, the C:N of the dairy and swine manures may be enhanced by adding food processing residues such as potato waste with a C:N ratio of 28:1, or crop residues, such as oat straw with a C:N ratio of 48:1 (EPA, 2012).

Foaming Nuisance and Control Strategy

Foaming in anaerobic digesters can cause a significant and wide spread problem. The most common cause of excessive foaming in anaerobic digesters is due to unstable operation resulting from the following:

- Shock loading or over loading to the digester

    o Organic overload to a digester generates more VFA than the methane producing bacteria can consume. Depending on the composition of the feed, shock loading or overloading can lead to higher rates of VFA and digester gas production and cause foaming problems (Massart et al, 2006). This can also happen if there is no blending tank or a blending tank with insufficient mixing for the various types of feed, such as sludge, FOG and food wastes before digester. A sudden change in feed composition or feeding mix can also upset the operational condition.

- Change in mixing Patten or regimen inside the digester

    o Inadequate mixing, excessive mixing or sudden power outage or mixing failure can contribute to this condition.

- Inadequate or excessive heating

    o This can result from the inability to maintain a stable temperature inside the digester causing temperature and density gradients resulting foaming problem.

- Higher ratio of biological sludge and foam causing micro organisms.

    o Higher biological sludge content inside the digester introduce higher amount of foam causing microorganism (*Nocardia, Microthrix parvicella* etc.) to the digester and increase the probability of foaming.

In order to control foaming the following preventive and control strategy can be adapted:

- Daily variations in volatile solids loading (organic load) to the digesters should be limited to 5 –10 percent. Better performance can obtained from feeding digester as continuously as possible, especially with FOG, food and high-strength organic wastes.
- Digester over-mixing, under mixing or improper mixing is a wide-spread concern and can contributes to foam production. Co-digestion with high strength waste can further add to this problem. That is why digester mixing system should be selected carefully considering the optimum mixing requirements as well as mixer performance.
- Digester should be design with additional head space and overflow capacity keeping the occasional foaming in mind.
- Further prevention can be added by installing foam sensor, surface discharge or removal option of foam, foam trap on gas line and protection to the pressure release valves
- Once foaming incident occurs it can be controlled by physicallly breaking up the foam using sprays or adding defoamants or foam suppressant chemicals in the feed.

Chapter 6

Biogas Cleaning, Upgrading and Utilization

Biogas can be used in many applications as an alternative to fossil fuel. The most common uses for biogas are electricity production and heat generation. But with the new and emerging biogas upgrading technologies, it can be converted to renewable natural gas and compressed renewable natural gas as well to be used as vehicle fuels. In addition, the versatile uses of biogas allow facilities to provide multiple end-uses at the same location; thus allow biogas to play a larger role in the renewable energy sector.

Depending on the end use, different biogas cleaning and upgrading steps are necessary. For example, when it is intended to use as a vehicle fuel or grid injection, the gas needs to be upgraded to increase its energy value. Although upgrading adds extra cost to the biogas production, the main advantage of upgrading is that; it can be used as an alternative to natural gas. This helps replacing the fossil fuel and reducing the green house gas emission at the same time.

Composition of Biogas

Biogas is composed mainly of methane (CH_4) and carbon dioxide (CO_2). Various other gases are also present in a low concentration. The methane content of biogas is very feedstock specific and varies with different composition of feedstocks. Typical ranges of different components of raw biogas as generated from anaerobic digestion are as follows (Source: American Biogas Council):

- Methane (CH_4):   55% to 60%
- Carbon dioxide (CO_2):   40% to 45%
- Nitrogen (N_2):   0.4 to 1.2 %
- Oxygen (O_2):   0.0 to 0.4%
- Hydrogen Sulfide (H_2S):   0.02 to 0.4%

Other trace gases or contaminants like ammonia, siloxanes and hydrocarbons are also presence in very low concentrations. Other contaminants also include water vapor, non-gaseous particulate and oil

Biogas Cleaning Technologies

The primary objective of biogas cleaning is to remove biogas components like H_2S, water vapour, NH_3, particles, etc. These components not only cause environmental hazard and processing problems but also dilute the energy density of biogas. The major contaminants of biogas and their related removal technologies are discussed as follows:

Table 6.1: Biogas Contaminants and Related Cleaning Technologies

Contaminants	Removal Technologies
Water Vapor: Biogas generated from anaerobic digestion is usually saturated with water vapor. Water vapour may condense into water or ice and thus result in corrosion and clogging issues. Most biogas utilization processes require relatively dry gas, so removal of water vapor is required.	1. **Passive cooling**: Biogas pipe line is run though underground for a short period of time. Water condenses from the biogas as it cool down. The condensate either discharges to sewer or recycle back. 2. **Refrigeration and Pressurization**: Heat exchangers can be used to cool down the biogas so that the water vapour gets condensed. Biogas can be further pressurized to dry it more. 3. **Absorption**: Biogas can be passed through drying medium like glycol, hygroscopic salts, silica gel, aluminum oxide etc. to absorb water. These drying medium can be regenerated by drying them at high temperature and sometime at high pressure as well. Eventually the drying media has to be replaced.
Hydrogen Sulphide (H2S): H2S is a toxic and corrosive gas and its concentration in raw biogas may vary based on the feedstock. H2S in biogas has to be reduced to harmless	1. **Water scrubbing**: Biogas is feed from the opposite direction of water flow to create a solution of H2S in water. The water can be regenerated and scrubbing water discharged can be reduced. 2. **Activated Carbon**: Biogas is fed through an activated carbon filter which removed sulphides by adsorption.

Contaminants	Removal Technologies
level to protect the downstream processes and equipments as well as toxicity to human health.	Activated carbon media can be regenerated. 3. **Iron Hydroxide or Oxide:** Biogas is led through a media composed of woodchips and iron oxide or hydroxide. H2S is removed as iron oxides react with sulfides (H2S) to produce iron sulfide. Bed can be regenerated several times before requiring replacement. 4. **Biofiltration:** Biofiltration uses microbes living on a packed medium to remove sulfides. Sulfides in the biogas get absorbed into a liquid film and are then metabolized by the microbial cells. It is available as above grade packed towers or below grade systems, filled with natural media like wood chips or peat moss.
Ammonia: Ammonia can cause corrosion to the downstream process equipments. It can also form nitrogen oxides (NO_x) from the combustion of biogas.	Ammonia is soluble in water, so it can be removed using the following two methods: 1. **Water Scrubbing** 2. **Refrigerated water vapour removal methods**
Particles: Sometimes dust and oil particles from the compressors may be present in the biogas.	1. **Filter:** The particles in the biogas can be removed by using filter of 2 to 5µm size. These filters need to be replaced in regular interval as part of maintenance.
Siloxanes: The presence of siloxanes in	1. **Silica gel:** Silica gel has good potential of removing of certain siloxane compounds from biogas. It has High adsorption capacity.

Contaminants	Removal Technologies
biogas cause abrasive siloxane deposits on equipments and reduce their life significantly.	2. **Cooling:** Siloxanes can be removed from the cooled gas with condensation water. 3. **Activated carbon:** Activated carbon removes siloxane from biogas via adsorption.
Halogenated hydrocarbons: Halogens can cause corrosion to mechanical parts of the plant. They can also form dioxins and furans during combustion of biogas.	1. **Activated carbon:** Activated carbon can be used to remove halogenated hydrocarbons.

Data Source:

- Anaerobic Digestion Guideline by the Ministry of Environment (MOE) of British Columbia (BC), Canada.
- Petersson and Wellinger, 2009
- Biogas Processing for Utilities, American Biogas Council

Biogas Upgrading Technologies

The main purpose of biogas upgrading is to increase the methane concentration of the biogas by removing contaminants that dilute its energy density, primarily carbon dioxide. Upgrading of biogas to a higher CH_4 concentration (>95%) opens up the possibility of biogas utilization as an alternative to natural gas. Upgrading the biogas to natural gas quality also makes another alternate application feasible and that is to use as a fuel for vehicles. However, biogas upgrading adds additional costs and overall plant economy should be considered prior to selecting the technologies.

Currently there are several full scale biogas upgrading technologies available. A brief description of these technologies is given below:

Absorption/Scrubbing

The principal behind absorption/scrubbing techniques is to absorb carbon dioxide from the biogas using a liquid (water, chemical or organic solvent) that passes through a packed column in a counter flow pattern form the biogas. Carbon dioxide has more solubility than methane. So the liquid exits the column with a higher concentration of carbon dioxide and the gas with a higher concentration of methane. Absorption technology can be categorized into three types based on the absorbents it uses (Petersson and Wellinger, 2009):

1. Water Scrubbing,
2. Organic Physical Scrubbing
3. Chemical Scrubbing

Water Scrubbing:

Biogas and water are passed in a counter flow pattern through a packed column filled with media to enhance contact between those two. Carbon dioxide gets dissolved in the water at a higher rate, so the gas that leaves the chamber will have a higher concentration of methane. The scrubbing water, containing mainly carbon dioxide and some methane is then taken to a flash tank where the gases are released. The gases from the flash tank can either be recycled back to the raw gas inlet or released to the atmosphere through a filter to prevent potential release of hydrogen sulfide (H_2S). Water scrubbing is the most common upgrading technique and is commercially available from different suppliers.

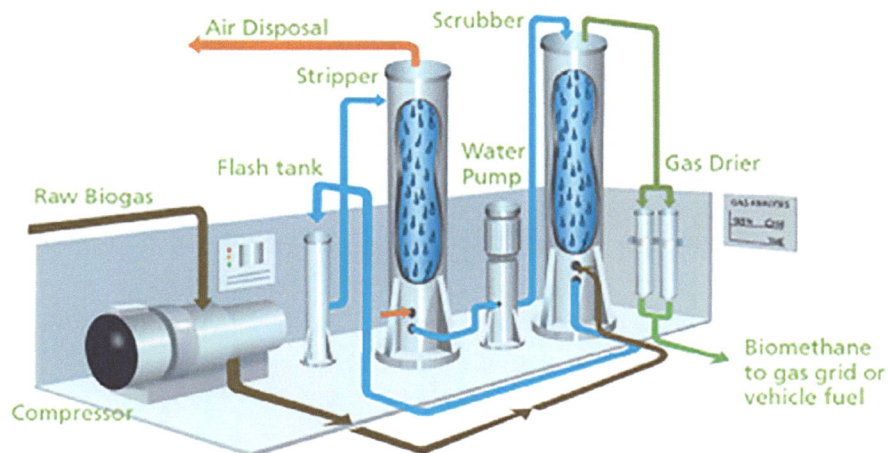

Figure 6.1: A Schematic Diagram of Water Scrubbing
(Source: Patrick Serfass, Biogas Processing for Utilities, American Biogas Council)

Organic physical scrubbing

Organic physical scrubbing uses an organic solvent such as polyethylene glycol instead of water for absorbing carbon dioxide. The organic solvent can be regenerated by heating and depressurizing. Other pollutants in biogas like hydrogen sulphide, water, oxygen and nitrogen may also be removed together with carbon dioxide, although more often they are removed prior to upgrading (Petersson and Wellinger, 2009). Selexol®, Genosorb® etc. are examples of trade names for solvents used in organic physical scrubbing.

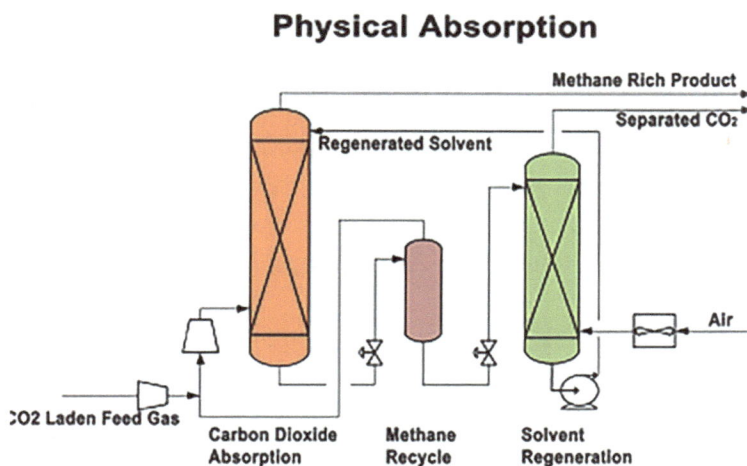

Physical Absorption

Methane Rich Product

Separated CO₂

Regenerated Solvent

Air

CO2 Laden Feed Gas

Carbon Dioxide Absorption · **Methane Recycle** · **Solvent Regeneration**

Figure 6.2: A Schematic Diagram of Organic Physical Scrubbing
(Source: Gail Richardson, Energy Vision, 2010)

Chemical scrubbing

Chemical scrubbers use chemical solution like mono ethanol amine (MEA), di–methyl ethanol amine (DMEA) etc. Amine solution reacts with carbon dioxide and since this chemical reaction is strongly selective, the methane loss may be as low as <0.1% (Petersson and Wellinger, 2009). The liquid solution is regenerated by heating. It is recommended to remove H_2S from the biogas before sending it to in the amine scrubber.

Pressure Swing Adsorption

Pressure swing adsorption (PSA) uses highly adsorbent materials like zeolites (crystalline polymers), carbon molecular sieves or activated carbon to selectively bind and release one or more contaminant gases from the biogas. At high pressure, selected contaminants get trapped in the adsorbent medium and later they are released at low pressure. Depending on the type of adsorbent and operating pressure used, different contaminants like CO_2, O_2 and N_2 can be adsorbed leaving methane gas to pass through. As the gas stream is subjected to "swings" from high to low pressure, the name pressure swing adsorption was given. The removal is done with multiple vessels to produce a continuous flow system. Liquid water and H_2S must be removed before the process but biogas containing water vapor is not an issue for the PSA.

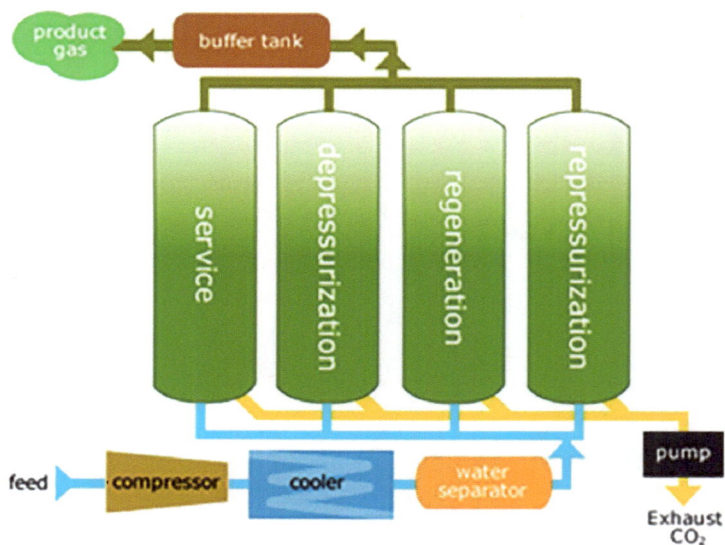

Figure 6.3: A Schematic Diagram of Pressure Swing Adsorption
(Source: Patrick Serfass, Biogas Processing for Utilities, American Biogas Council)

Membrane Separation

Membranes material used for biogas upgrading, usually hollow fibre membranes; are permeable to undesirable gases like carbon dioxide, water and ammonia while rejecting methane. As a result a high methane content gas generates from the "retentate". Usually the process is performed in multiple stages and the two stage systems are the most common one. Usually biogas needs to go through desulphurization and drying steps before going into membrane separation for upgrading.

Two Stage Membrane

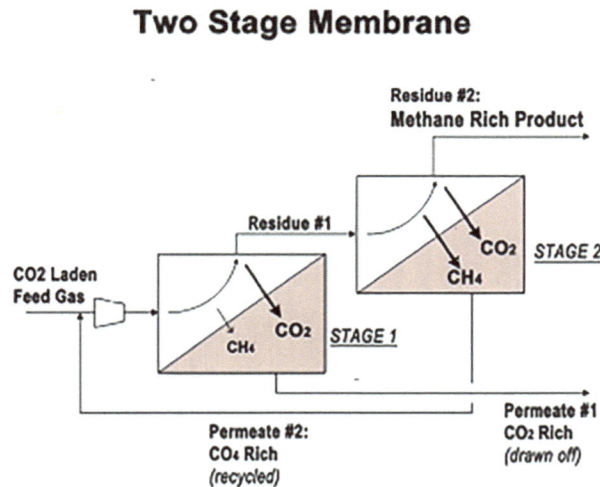

Figure 6.4: A Schematic Diagram of Two Stage Membrane separations
(Source: Gail Richardson, Energy Vision, 2010)

Cryogenic Separation/Distillation

Cryogenic separation or distillation uses the different condensation temperatures of the gases to separate them from each other. Different contaminants gases of biogas such as CO_2, H_2S etc. liquefy at different temperature–pressure zones than the main desirable component CH_4 and this is the principal that used in the cryogenic separation technique. Initially the process starts with compressing the biogas. Then several different heat exchanger and compressor steps are used to gradually cool the biogas to a lower temperature (near –100 ^{0}C). Finally, at the distillation column, biogas contaminants mainly H_2S and CO_2 get separated from CH_4 in the form of a liquid fraction. The schematic diagram of a cryogenic separation is shown in the following figure (source: http://students.chem.tue.nl/ifp24/techn_cryo.htm).

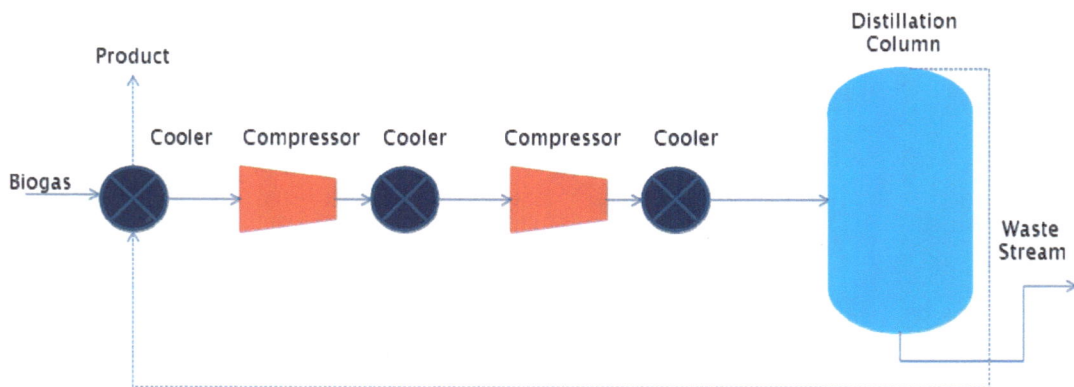

Figure 6.5: Schematic Diagram of a Cryogenic Separation

The main advantage of cryogenic separation is the high purity of the upgraded biogas. But the main disadvantage of cryogenic separation is that it requires higher process equipments, mainly compressors, turbines and heat exchangers. This increases process complicacy as well as capital cost.

Comparison of Different Upgrading Technologies

The comparisons of different biogas upgrading technologies are shown in the following table as presented in Petersson and Wellinger, 2009.

Table 6.2: Comparison of Different Upgrading Technologies
(Data source: Urban et al. 2008, as presented in Petersson and Wellinger, 2009)

Parameter	PSA	Water Scrubbing	Organic Physical Scrubbing	Chemical Scrubbing
Pre-cleaning needed [a]	Yes	No	No	Yes
Working pressure (bar)	4–7	4–7	4–7	No pressure
Methane loss [b]	< 3 % / 6 –10% [f]	< 1 % /< 2% [g]	2– 4%	< 0.1%
Methane content in upgraded gas [c]	> 96%	> 97%	> 96 %	> 99%
Electricity consumption [d] (kWh/Nm3)	0.25	< 0.25	0.24–0.33	< 0.15
Heat requirement (°C)	No	No	55–80	160
Controllability compared to nominal load	+/– 10-15%	50 –100%	10 –100%	50 –100%
References [e]	> 20	> 20	2	3

a. Refers to raw biogas with less than 500 mg/m³ of H_2S. For higher concentrations, pre-cleaning is recommended also for the other techniques.
b. The methane loss is dependent on operating conditions. The figures given here refer to figures guaranteed by the manufacturers or provided by operators.
c. The quality of biomethane is a function of operational parameters. Figures given refer to figures guaranteed by the manufacturers or provided by operators, based on air-free biogas.
d. Given in kWh/Nm³ of raw biogas, compressed to 7 bar (g).
e. Number of references reviewed. Some are pilot plants.
f. <3 % CarboTech, / 6-10 % QuestAir.
g. < 1 % Malmberg / <2 % Flotech.

Biogas Upgrading Plant/Technology Suppliers

List of different biogas upgrading plant/technology suppliers are listed in Table 6.3 as per IEA bioenergy website. Data was extracted from the site (http://www.iea-biogas.net/plant-list.html) on August 14, 2014.

Table 6.3: List of Different Biogas Upgrading Plant/Technology Suppliers

Technology Type	Company and Home Page
Manufacturers of PSA units	• Acrona-systems, www.acrona-systems.com • Carbotech, www.carbotech.info • Cirmac, www.cirmac.com • ETW Energietechnik, www.etw-energy.com • Guild, www.moleculargate.com • Mahler, www.mahler-ags.com • Strabag, www.strabag-umweltanlagen.com • Sysadvance, www.sysadvance.com • Xebec, www.xebecinc.com
Manufacturers of water scrubbing units	• DMT, www.dmt-et.nl • Econet, www.econetgroup.se • Greenlane Biogas, www.greenlanebiogas.com • Malmberg Water, www.malmberg.se • RosRoca, wwwrosroca.com

Technology Type	Company and Home Page
Manufacturers of chemical scrubbing units	• Arol Energy, www.arol-energy.com • BIS E.M.S. GmbH, www.ems-clp.de • Cirmac, www.cirmac.com • Energy & waste technologies, www.ewtech-ing.com • Hera, www.heracleantech.com • MT-Biomethan, www.mt-biomethan.com • Purac Puregas, www.lackebywater.se • Strabag, www.strabag-umweltanlagen.com
Manufacturers of organic physical scrubbing units	• HAASE Energietechnik, www.haase.de • Schwelm Anlagentechnik, www.schwelm-at.de
Manufacturers of membrane units	• Air Liquide, www.airliquide.com • Arol Energy, www.arol-energy.com • BebraBiogas, www.bebra-biogas.com • Biogast, www.biogast.nl • Cirmac, www.cirmac.com • DMT, www.dmt-et.nl • Eisenmann, www.eisenmann.com • EnviTec Biogas, www.envitec-biogas.com • Gastechnik Himmel, www.gt-himmel.com • Haffmans, www.haffmans.nl • Mainsite Technologies, www.mainsite-technologies.de • Memfoact, www.memfoact.no • MT-Biomethan, www.mt-biomethan.com • Prodeval, www.prodeval.eu

Technology Type	Company and Home Page
Manufacturers of cryogenic units	• Acrion Technologies, www.acrion.com • Air Liquide, www.airliquideadvancedtechnologies.com • Cryostar, www.cryostar.com • FirmGreen, www.firmgreen.com • Gas treatment Services, www.gastreatmentservices.com • Gasrec, www.gasrec.co.uk • Hamworthy, www.hamworthy.com • Prometheus Energy, www.prometheusenergy.com • Terracastus Technologies, www.terracastus.com
Manufacturers with special focus on small scale biogas upgrading	• Biosling, www.biosling.se • Biofrigas, www.biofrigas.se • Metener, www.metener.fi • Neo-Zeo, www.neo-zeo.com • Sysadvance, www.sysadvance.com
Gas entry unit system Integrators	• Orbital, www.orbital-uk.com

Biogas Utilization

Electricity and heat are still the most convenient application for many biogas plants. Most biogas-to-electricity producing plants usually use the waste heat from the IC engine/ microturbine to heat the digester. Beneficial use of the heat produced during electricity generation is called combined heat and power generation or cogeneration, and this increases the overall energy efficiency of the systems.

Recently, the advancement in technology has created other beneficial usage of biogas and that is to produce renewable natural gas (RNG) and compressed renewable natural gas. The renewable natural gas can be injected to existing gas grid to be utilized in various residential commercial and industrial usages. RNG must meet some specific minimum or maximum levels for certain gases, moisture, and other constituents in order to be added into the existing gas grid and it must be pressurized before injection. But once RNG is added to the pipeline it is considered to be no different from natural gas.

Figure 6.6: Injection System Used to Pump RNG into a Natural Gas Line at a Farm-Based Biogas System
(Source: /www.omafra.gov.on.ca)

In addition, RNG can be compressed to be used as a transportation fuel. Compressed RNG can also be stored in pressurized containers to be used on-site or at remote locations. Compressed RNG can also be taken to vehicle fuelling stations using trailer-mounted pressurized containers.

The following schematic shows some beneficial end use of biogas along with their cleaning and upgrading requirements:

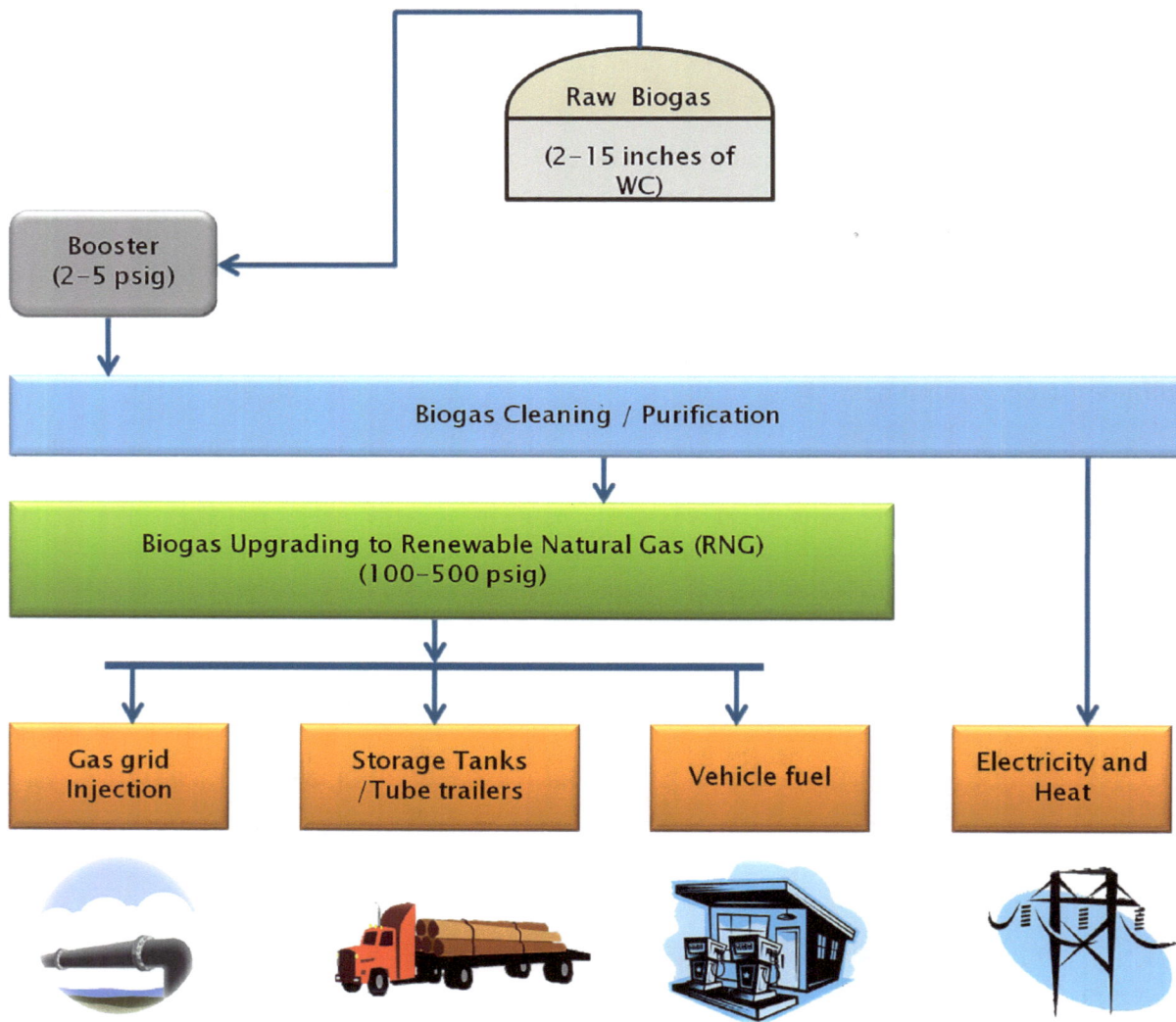

Figure 6.7: Some Beneficial End Use of Biogas along with their Cleaning and Upgrading Requirements.
(Data source: American Biogas Council)

Chapter 7

Digestate Treatment and Resource Recovery

The nutrient rich by product of anaerobic digestion process is known as digestate. Digestate usually contains all the nitrogen, phosphorus and potassium present in the original feedstocks. As a result, it has been used as fertilizer and soil conditioner historically. But currently other usages of digestate are also being investigated to reduce land-based application and associated transportation cost. Recently some promising digestate enhancing and recovery options have been evolved and those are going to be discussed in this chapter. But ultimately the digestate treatment and recovery option will depend on the overall plant economy and rate of return of the final products. So, before adopting any recovery option or treatment train the following points need to be considered:

- Is the chosen treatment train or recovery route most economically viable one considering market demand, production cost and rate of return?
- What is the state of the chosen technology and what is the level of risk associated with its performance?
- Does the end product or its final usage meet all the local regulatory compliance and requirements?

Digestate Treatment and Enhancement Options

Digestate contains some essential plant nutrients such as Phosphorous (P), Nitrogen (N), Potassium (P) and Sulfur (S). However, they may also contain contaminants such as heavy metals, toxic chemicals, pharmaceuticals, infectious microorganisms, etc. depending on the feedstock used. That is why digested treatment should be chosen based on its physical, chemical and biological properties as well as desired end use or disposal options.

The current and evolving digestate treatment and enhancement options can be categorized in to the following groups:

- Physical Treatment and Enhancement Options
- Thermal Treatment and Enhancement Options

- Conversion Treatment and Enhancement Options
- Biological Treatment and Enhancement Options
- Chemical Treatment and Enhancement Options

These categories are briefly described in the following sections

Physical Treatment and Enhancement Options

In a wet anaerobic digestion process digestate is basically in a liquid state with around 4–10% of solids. To convert this digestate as a valuable fertilizer, nutrients need to be concentrated or in the other words, water in the digestate needs to be reduced. This is usually done through various separation technologies like thickening, dewatering, filtering etc., that separate the fiber or solid part of the digestate from the liquid part. The separated liquid part can also be used as process water or recycle water after receiving further treatment. Several separation technologies are discussed as follows:

Thickening

Thickening is usually done at the pre–treatment stage to reduce digester volume as well as subsequent storage. This is a partial separation process for solid–liquid and the final liquor has usually 5–10% solids based on the technology adopted. Thickening is generally accomplished by physical process like co–settling, gravity settling, flotation, centrifugation, gravity belt and rotary drum thickener technology (Metcalf and Eddy, 2003). Sometime different coagulants are also added to increase solid capture and process efficiency.

Dewatering

Dewatering is applied when a higher percentage of solids are desired in the final separated product. Generally dewatering technology can achieve greater than 20% of solids in the separated cake of the digestate. Some of the advantages of dewatering include:

- Lower cost of trucking and transportation of dewatered digestate to the final disposal site.
- Dewatered digestate are easier to handle than the thickened one because of their solid nature.
- Dewatering makes the subsequent treatment like incineration, composting etc., cost effective.

The commonly used dewatering technologies are solid-bowl centrifuges, belt filter presses, recessed-plate filter presses etc. The selection of technology is usually decided based on the characteristics of the digestate, the desired characteristic of the dewatered products as well as the economy and space available in the treatment plant. Different coagulants are also added during the dewatering process to improve solid capture and operability of the process.

The liquid generated from dewatering process contain high amount of ammonia and potassium and required further treatment prior to disposal. This liquor can also be used as process water for feedstock dilution or recycle water based on its quality and reuse requirements.

Membrane Filtration

Membrane acts as a physical barrier or molecular sieve and allows the water to pass through while retaining the solids. Different type of membrane products and technologies are available as a separation tool like microfiltration (MF), ultra filtration (UF) and Reverse Osmosis membrane. But due to their small pore sizes, membranes are very susceptible to fouling and they are usually used to treat the separated liquor from the digestate to purify them further. High quality product water, suitable for reuse, can be achieved through membrane separation. The concentrate from membrane separation process can be used as fertilizer.

Thermal Treatment and Enhancement Options

In thermal treatment, thermal energy or heat is being used to dry up the digestate. This can be achieved by using various types of dryer or evaporators.

Different type of drying bed and drying technologies are available (Metcalf and Eddy, 2003); such as:

- Conventional sand drying bed
- Paved drying bed
- Artificial media drying beds
- Vacuum assisted drying beds
- Solar drying beds
- Rotary sludge dryer
- Fluidized bed dyer
- Evaporator

The thermally dried product has a greatly reduced water content and volume. As a result it is very convenient for handling, storing and trucking. The dried product can be palletized to further facilitate storage, transportation and use as a fertilizer.

The evaporated fraction resulted from thermal drying needs further treatment before disposal.

Figure 7.1: HUBER Solar Active Dryer SRT
(Source: http://www.huber.de)

Integrated Dewatering and Drying System

Current advancement of technology has allowed performing both dewatering and drying in the same machine, generating a high-solid dried end product that is very suitable for handling and trucking. The J–VAP® dewatering and drying system by Evoqua (Siemens) water technologies is one of such integrated dewatering and drying system. As per manufacturer website, this system allows to obtain cake solids of over 95% in one-step operation. Vacuum and and hot water (150 to 180° F/65 to 82°C) are used in drying operation. The system has also option to select desired moisture content up to 99% solids that allows customizing the drying process and the quality of the end product (http://www.evoqua.com/en/products/sludge_biosolids_processing/filter_press/Pages/dewatering_systems_product_jvap.aspx).

As per the manufacturer, other benefits associated with this technology are:

- Automated operation
- Significant weight/volume reduction
- Dry sludge without air discharge permits
- Pathogen reduction to meet Class A
- Eliminate multiple equipment and handling
- Reduce disposal costs

Figure 7.2: The J-VAP® Dewatering and Drying System
(Source: http://www.evoqua.com)

Conversion Treatment and Enhancement Options

In the conversion treatment, digestate is converted or break down into different end-products that can be used in alternate applications rather than the conventional land based one. Most of the conversion technologies require dewatered sludge with low moisture content and preferably in palletized form.

Incineration

In incineration, digestate is totally converted to oxidized end products mainly carbon dioxide, water and ash. The ash generated in the process can be used as a construction material for road or concrete. Phosphorus can also be recovered from ash by acid leaching (WRAP, 2012). The advantages of incineration are maximum volume reduction, pathogen and toxic compound destruction as well as resource recovery potential. But the high capital and operating cost; adverse environmental effects associated with the process emission and potential hazard of the generated residual make it a less preferable technology.

Gasification

In gasification process, partial combustion of organic matter is conducted to produce an end product called syngas (synthesis gas). Syngas is mainly a mixture of carbon monoxide and hydrogen and can be used to produce energy. The process prefers a dry palletised form and low moisture content in the feed (WRAP, 2012).

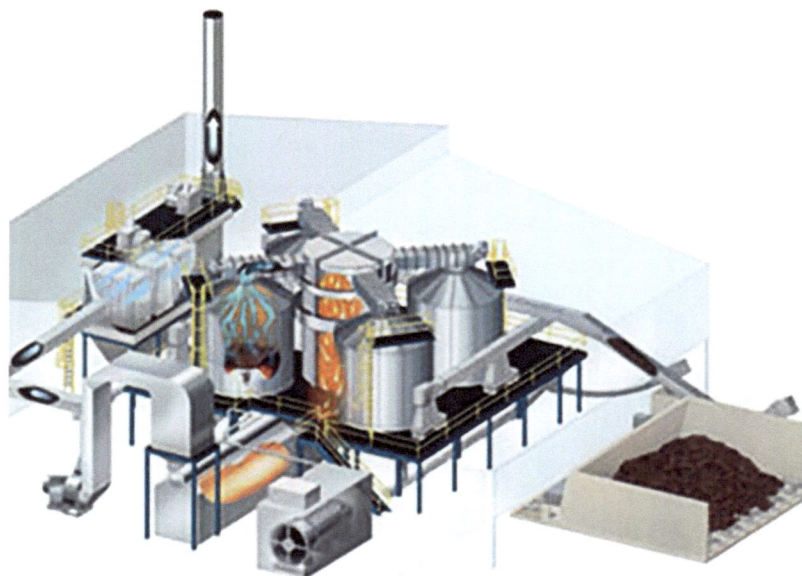

Figure 7.3: A Gasification System by Nexterra
(Source: www.nexterra.ca)

Wet Air Oxidation

In the wet air oxidation (WAO) process digestate is oxidised within the liquid phase, rather than in the gaseous phase, in contrast to the other conversion processes. This process is achieved at elevated temperatures and high pressure to prevent evaporation. The end products are a mineral sludge, a liquid effluent and off gasses (WRAP, 2012). There are usually no pre-treatment required for this process.

Pyrolysis

Pyrolysis process breaks down the organics of digestate in the presence of heat and in the absence of oxygen. Similar to incineration and gasification, pyrolysis process prefers palletized form and low moisture content in feed. The end products of pyrolysis are char and syngas. While syngas can be used as an energy source, char can be used as a soil amendment.

Figure 7.4: A Pyrolysis System by 3R Agrocarbon
(http://www.3ragrocarbon.com/)

Biological Treatment and Enhancement Options

In biological treatments and enhancement options for digestate, microbial species have been utilized for reducing organic content as well as for producing useful end products like fertilizer and biofuel. Some of them are discussed as follows:

Composting

Composting utilizes a variety of microbial communities including bacteria, actinomycetes and fungi to stabilize the digestate. In composting around 20 to 30 percent of solids are converted to carbon dioxide and water. Composting can be accomplished both aerobically and anaerobically although partial aerobic condition is the most common one in practice (Metcalf and Eddy, 2003).

Sometimes amendments materials like saw dust, straw etc. are added with the feed to aid aeration and reduce mixture weight and moisture content. Moreover, adding a bulking agent is also common to provide structural support and to increase porosity of the mix. The most commonly used bulking agents are wood chips. Some design consideration for aerobic sludge composting are as follows (Metcalf and Eddy, 2003):

- The Initial C/N ratio should be in the range of 20:1 to 35:1 by weight
- The volatile solid of the mix should be greater than 30% of the total solids
- pH of the mixture should be in the range of 6–9
- The temperature should be in the 50–60 $^{\circ}$C for optimum result

Currently, different types of composting methods are in practice, such as aerated static pile, windrow, or in vessel composting systems. At the end, Composting operation and its subsequent use should match with the local regulations to ensure public health and environmental safety.

Figure 7.5: Multi Bin Composting, In-vessel Composting and Hoop-structure Composting
(Source: http://www.gov.mb.ca)

Reed beds

Reed bed is a passive composting system that dewaters and stabilizes the digestate over a long period of time. Typically the whole digestate is feed into the bed of reeds and then it gets treated by the microorganism at the root of the plants. The water gets removed over time via collection basin and evapo-transpiration. Reed bed is similar in appearance of subsurface flow constructed wetland.

Figure 7.6: A Reed Bed System
(Source: http://www.armreedbeds.co.uk/)

Biological Oxidation

Biological oxidations are various biological wastewater treatment processes that can employ to treat either the whole digestate or the liquid portion of the digestate. Process selection depends both on the digestate solid content as well as solid tolerance capability of the process. Some of such biological treatment processes are Sequencing Batch Reactors (SBR), Membrane Bioreactors (MBR) etc. These processes use bacterial population that oxidized BOD and ammonia in the presence of oxygen. The effluent resulted from these process are in high quality and reusable in many cases. The biological biomass can be returned to the digester.

Figure 7.7: A Full Scale Membrane Based System to Treat Digestate by bkt21 in Netherlands
(Source: http://www.bkt21.com/digestate-treatment-and-nutrient-recovery/)

Biofuel Production

Digestate has the potential to be used as feedstock for biofuel production. The liquid part of the digestate can be used to grow algal population and produced algae later can be used for biofuel production. Lipids from the algae can be separated and converted into biofuel, and the remaining algal biomass can either be sent back to the digester or sold as animal feed. The separated water from algal population can be used as process water or for other irrigation purposes (WRAP, 2012). Several such techniques are currently being developed.

Figure 7.8: A view of an Algal Bioreactor for Biofuel Production
(Source: WRAP, 2012)

Chemical Treatment and Enhancement Options

Struvite Production

Phosphorus and ammonia can be recovered from the digestate or its centrate in the form of Struvite ($MgNH_4PO_4 \cdot 6H_2O$), a solid compound made of magnesium, ammonium and phosphate. Struvite can be used as a fertilizer and because of its slow release property; it stays in the soil longer rather than washed off by the water.

The struvite precipitation is usually carried out in a fluidized bed type of reactor. The reaction is highly pH dependant with optimum value being at around pH 10. Usually magnesium is added to the process and pH is artificially elevated to create the optimum environment. pH can be increased by dosing with sodium or magnesium hydroxide and Magnesium ion concentration can be increased by adding magnesium hydroxide or chloride. But adding magnesium hydroxide is advantageous because it increase both pH and magnesium ion in the solution (WRAP, 2012). Struvite contains equimolar amounts of ammonium and phosphate and digested usually contains more ammonium than phosphate. So the resulting effluent may require more treatment to recover/remove ammonia before disposal.

Ostara, a Canadian nutrient recovery company has developed a struvite production technology, known as Pearl® Nutrient Recovery Process that has several full scale installations in Canada and USA (http://www.ostara.com/nutrient-management-solutions). This technology is based on controlled chemical precipitation in a fluidized bed reactor that recovers struvite in the crystallized form. Magnesium chloride and sodium hydroxide are added when necessary. Struvite seeds form and grow in the process until they reach the desired size of 1.0 mm to 3.5 mm. As per manufacturer website, up to 90 per cent of the phosphorus and 40 per cent of the ammonia load can be removed from sludge dewatering liquid in a municipal wastewater treatment plant by using this process. The resulting product is marketed as a commercial fertilizer called Crystal Green®.

Ostara's commercial-scale nutrient recovery systems are currently operating at several wastewater treatment plants in North America and Europe (http://www.ostara.com/nutrient-management-solutions/installations).

Figure 7.9: Struvite Recovery at Durham Wastewater Treatment Plant, Tigard, Oregon, USA
(Source: http://www.hdrinc.com)

Ammonia Stripping

Usually ammonia stripping is conducted in a packed column tower. Ammonia reacts with water to form ammonium hydroxide. In ammonia stripping of digestate–centrate, pH of the solution is increased until it reaches 10.8 to 11.5 (EPA, 2000). At increased temperature and pH, ammonium hydroxide converts to ammonia gas. Then air or steam is passed through the tower in a counter flow pattern to absorb the ammonia. This stripping gas is then passed through an acid scrubbing tower, operating with either sulfuric or nitric acid. The ammonia comes out from the scrubber as ammonium sulphate or nitrate based on the acid used for scrubbing. The process air/steam can be reused (WRAP, 2012).

Recent development suggested that biogas can be used instead of air to strip the ammonia. Walker et al., (2011), suggested different scenarios of ammonia removal from food waste digestate using biogas stripping. The results of the modelling done by the authors showed that ammonia removal in an integrated process was achievable in all of scenarios considered. In addition, they suggested that in situ ammonia stripping is a promising treatment option for decreasing in-digester ammonia concentrations, especially at a high organic loading rate.

Ammonia can also be recovered via ion exchange process where digestate is fed into a packed bed of adsorbent like zeolites, clays or resins. There ammonium gets selectively adsorbed by ion exchange. Once saturated the column is taken off-line and regenerated, recovering the ammonium (WRAP, 2012).

Alkaline stabilization

Alkaline stabilization is a method of eliminating nuisance condition of digestate using alkaline material like lime. In lime stabilization process the pH of the digestate is raised to 12 or higher to create an environment that halts or slowdowns further microbial reactions. As a result, the digestate will not putrefy, create odor or create a health hazard as long as the high pH condition remains. This process can also inactivate microorganisms and pathogens present in the digestate (Metcalf and Eddy, 2003).

Table 7.1: A Summary of Different Treatment and Enhancement Option for Digestate.

Treatment and Enhancement Options	Methods
Physical Treatment and Enhancement Options	• Thickening (Belt, Centrifuge) • Dewatering (Belt Press, Centrifuge) • Membrane Filtration (Microfiltration, Ultra filtration and Reverse Osmosis)
Thermal Treatment and Enhancement Options	• Drying (Rotary Drying, Belt Dryer, Drying Bed) • Evaporation (Heat Exchanger)
Conversion Treatment and Enhancement Options	• Incineration • Gasification • Wet Air Oxidation • Pyrolysis
Biological Treatment and Enhancement Options	• Composting • Reed Beds • Biological Oxidation • Biofuel Production
Chemical Treatment and Enhancement Options	• Struvite Precipitation • Ammonia Recovery (Stripping & Scrubbing, Ion Exchange, Membrane) • Alkaline Stabilization

Data Source: Enhancement and treatment of digestates from anaerobic digestion by WRAP November 2012, Written by Pell Frischmann Consultants Ltd.

Resource Recovery from Digestate

Biosolids and digestate have long been used to agricultural lands to take advantage of the nutrients and micronutrients of these products, with the aim to increase soil fertility or structure. But recently, a wide range of resource recovery options has evolved from the digestate beyond this conventional agricultural use. Some of the drivers behind this evolution are mainly search for renewable fuel sources, reducing greenhouse gases and reducing transportation cost to the suitable application sites. Also there are some disadvantages associated with land application such as high nitrogen content with potential of ammonia and nitrate pollution, high dilution requirements, needs for supplementary nutrient addition to create a balanced fertilizing feed etc.

At present, the focus is on complete nutrient recovery and creating value added products from digestate. In recent practices digestate is separated in to two different fractions– fibre and liquor. These fractions have different nutrient profile with different potential uses and potential markets.

The water content of digestate can be reduced through dewatering process. The solid resulted from dewatering process has reduced volume and thus has reduced trucking and fuel cost for its handling and disposal. Moreover, this solid can be used for composting to create a value added product for soil amendments. The composting product has fertilization benefits as well. The dewatered solid can be dried up further to make it more suitable for further energy recovery options like gasification and pyrolysis. These processes produce syngas that can be used as an alternative source of energy. In addition, pyrolysis produces char; which has potential to be used as soil amendments and fertilizer. Supplementary nutrient can also be added to the dried digestate product to create a balanced fertilizer.

The liquid generated from dewatering process known as centrate, can also be utilized further. This liquor is very high in nutrients like ammonia and phosphorus. Both of these nutrients can be precipitated from digestate liquor as struvite that can later be transformed into an organic fertilizer. As equal amounts of ammonia and phosphate are used in struvite production and digestate is relatively rich in ammonia, a significant quantity of ammonia can still be recovered through ammonia precipitation from the left-over liquor. Moreover, digestate liquor can be treated through any biological wastewater treatment process like MBR, SBR etc. which generates a high quality reusable effluent. The treated digested liquor can be used as process water or as irrigation water.

With various evolving resource recovery technologies from digestate, before selecting any treatment/ recovery technology, the overall plant economy should also be kept in mind. The end use of the final products and theirs existing market should also be considered.

Figure 7.10 shows different existing and emerging resource recovery options from digestate.

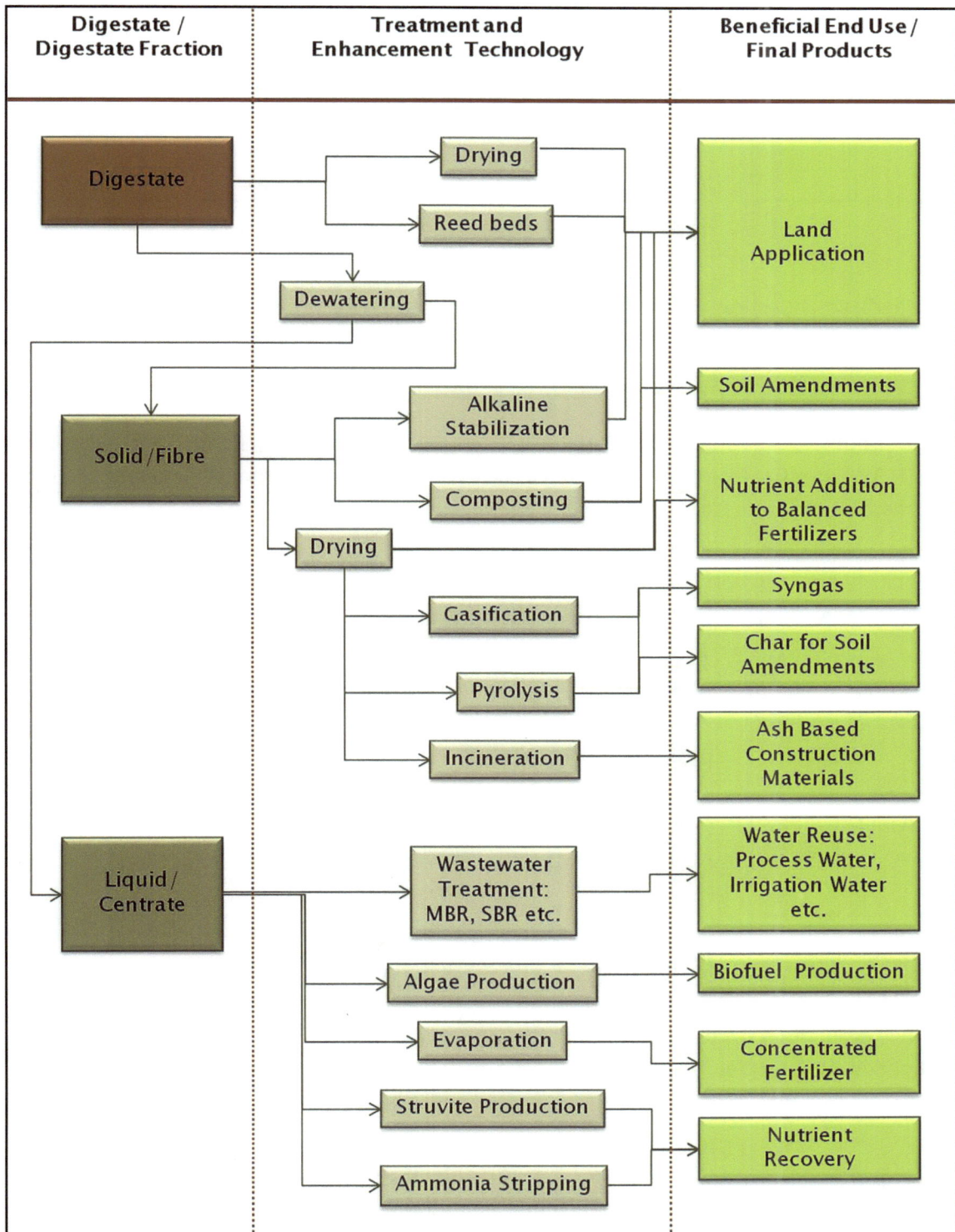

Figure 7.10: Various Resource Recovery Options from Digestate

Chapter 8

References

1. Anaerobic Digestion Guideline by the Ministry of Environment (MOE) of British Columbia (BC), Canada, Chapter-6, www.env.gov.bc.ca/epd/industrial/agriculture/pdf/adg/adg-chapter6.pdf.

2. Antognoni, S., Ragazzi, M., Rada E. C., Plank, R., Aichinger, P., Kuprian, M., Ebner, C. (2013); Potential Effects of Mechanical Pre-treatments on Methane Yield from Solid Waste Anaerobically Digested, http://pubs.sciepub.com/ijebb/1/1/4/index.html.

3. Black, A. (2012); Anaerobic Digestion Pre-Treatment system investigation, http://www.wrap.org.uk/sites/files/wrap/Green%20Farm%20Energy.pdf.

4. Braun, R. and Wellinger, A.; Potential of Co-digestion by IEA Bioenergy, Task-37, http://www.iea-biogas.net/files/daten-redaktion/download/publi-task37/Potential%20of%20Codigestion%20short%20Brosch221203.pdf.

5. Cavinato, C. (2011); Anaerobic Digestion Fundamentals, www.valorgas.soton.ac.uk/Pub_docs/JyU%20SS%202011/CC%201.pdf.

6. Chen, Y., Cheng, J. J., Creamer, K. S. (2008); Inhibition of Anaerobic Digestion Process: A Review, Bioresource Technology, 99: 4044-4064.

7. Davidsson. A., Lovstedt, C., Jansen, J. C., Gruvberger, C. and Aspegren, H. (2007); Co-Digestion of Grease Trap Sludge and Sewage Sludge, Waste Management, volume 28, issue 6:986-992.

8. Doháyos, M. Zábranská, Kutil, J. and Jeníček, P. (2004); Improvement of Anaerobic Digestion of Sludge, Water Science and Technology, 49 (10), 89-96.

9. EPA (2000); Biosolids Technology Fact Sheet, Centrifuge Thickening and Dewatering.

10. EPA (2000); Wastewater Technology Fact Sheet, Ammonia Stripping.

11. EPA (2006); Biosolids Technology Fact Sheet, multi-Stage Anaerobic Digestion.

12. EPA (2008); Anaerobic Digestion of Food Waste-Final Report Prepared by East Bay Municipal Utility District.

13. EPA (2011); Recovering Value from Waste, Anaerobic Digester System Basics.

14. EPA (2012); Increasing Anaerobic Digester Performance with Codigestion, www.epa.gov/agstar.

15. Erdal, U.G.; Soroushian, F.; Kitto, W.; Whitman, E.J. (2005); A Technology Review for Co-Digestion for Food Waste with Biosolids and/or Manure, Source: Proceedings of the Water Environment Federation, Residuals and Biosolids Management 2005, pp. 477-493(17).

16. Genesis Projects Corp (2007); Anaerobic Digestion- A Cost-effective and Environmentally Safe Option for the Disposal of Livestock Waste Tissue, www.iafbc.ca/.../livestock/documents/LWTI-11_Anaerobic%20Report.pdf

17. Gómez, P., Begoña, R., Pascual, A., Koch, K., Andrade, D. (2012); Influence of A Mechanical Substrate Pre-Treatment on The Anaerobic Digestion Process, http://cigr.ageng2012.org/images/fotosg/tabla_137_C1869.pdf

18. Heo, N.H., Park, S.C., Lee, J.S. and Kang, H., (2003); Solubilization of Waste Activated bSludge By Alkaline Pre-Treatment and Biochemical Methane Potential (BMP) Tests for Anaerobic Co-Digestion of Municipal Organic Waste. Water Science and Technology 48 (8), 211-219.

19. http://www.extension.org/pages/26617/feedstocks-for-biogas#.U9rMCJV0zIU; information extracted July 31, 2014.

20. Kaparaju, P. and Rintala, J. (2005): Anaerobic Co-Digestion of Potato Tuber and Its Industrial Byproducts with Pig Manure, Resources, Conservation and Recycling, 43: 175-188.

21. Knapp, J.S. and Howell, J.A. (1978); Treatment of Primary Sewage Sludge with Enzymes, Biotechnology and Bioengineering, 20, 1221-1234.

22. Kopp, J., Muller, J., Dichttl, N. and Schwedes, J. (1997); Anaerobic Digestion and Dewatering Characteristics of Mechanically Disintegrated Excess Sludge, Water Science and Technology, Vol-36, No. 11, 129-136.

23. Lehtom¨aki, A., Huttunen, S., Rintala, J.A. (2007); Laboratory Investigations on Co-Digestion of Energy Crops and Crop Residues with Cow Manure for Methane Production: Effect of Crop to Manure Ratio. Resources, Conservation and Recycling, 51: 591-609.

24. Li, Y.Y. and Noike, T. (1992); Upgrading of Anaerobic Digestion of Waste Activated Sludge by Thermal Pre-Treatment, Water Science and Technology, Vol-26, No. 3-4, 857-866.

25. Liu, X., Wang, W., Gao, X., Zhou, Y., Shen, R. (2012); Effect of Thermal Pretreatment on The Physical and Chemical Properties of Municipal Biomass Waste, Waste Management, 32(2):249-55.

26. MacFarlane, A. and Davis, W. (2011); Anaerobic Digestion Technologies, http://www.newmoa.org/solidwaste/cwm/webconf/adtech/AnaerobicDigestionWebinarJan2012.pdf

27. Massart, N., Bates, R., Corning, B., Neun, G. (2006); Design and Operational Considerations to Avoid Excessive Anaerobic Digester Foaming, WEFTEC.

28. Metcalf and Eddy (2003); Wastewater Engineering: Treatment and Reuse, 4th ed., McGraw-Hill, New York.

29. Moody, L., Burns, R., Wu-Haan, W., Spajiš, R. (2009); Use of Biochemical Methane Potential (Bmp) Assays for Predicting and Enhancing Anaerobic Digester Performance, 44th Croatian & 4th International Symposium on Agriculture, http://sa.agr.hr/pdf/2009/sa2009_a0707.pdf.

30. Mshandete, A., Björnsson, L., Kivaisi, A.K., Rubindamayugi, S.T. and Mattiasson, Bo. (2005); Enhancement of Anaerobic Batch Digestion of Sisal Pulp Waste by Mesophilic Anaerobic Pre-Treatment. Water Research 39, 1569-1575.

31. Münnich, K. (2008); Mechanical Biological Treatment of Msw– a Potential to Reduce the Impact on Environment, http://www.faber-ambra.com/docs/2008-11-MBT-Presentation-Leichtweiss-Institut-TU-Braunschweig.pdf.

32. Mukherjee, S.R. and Levine, A. D. (1992); Chemical Solubilization of Particulate Organics as a Pre-Treatment Approach. Water Science and Technology, Vol-26, No. 9-11, 2289-2292.

33. Müller, J.A. (2000); Pre-Treatment Processes for the Recycling and Reuse of Sewage Sludge. Water Science and Technology 42 (9), 167-174.

34. Navaratnasamy, M., Jones, J. and Partington, B. (2008); Biogas: Cleaning and Uses by Alberta, Agriculture and Rural Development.

35. Olvera, J. R. and Lopez A. L. (2012); Biogas Production from Anaerobic Treatment of Agro-Industrial Wastewater, Biogas, Dr. Sunil Kumar (Ed.), ISBN: 978-953-51-0204-5, InTech, Available from: http://www.intechopen.com/books/biogas/biogas-production-from-anaerobic-treatment-of-agro-industrialwastewater.

36. Owen, W., Stuckey, D., J. Healy, Jr., Young, L., McCarty, P. (1979); Bioassay for Monitoring Biochemical Methane Potential and Anaerobic Toxicity, Water Research, 13, 485-492.

37. Petersson, A. and Wellinger, A., (2009); Biogas Upgrading Technologies –Developments and Innovations, by IEA Bioenergy.

38. Rabinowitz, B. and Stephenson, R. (2006); Effect of Microsludge on Anaerobic Digester Performance and Residuals Dewatering at La County's JWPCP, WEFTEC, http://www.environmental-expert.com/Files/5306/articles/8537/039.pdf.

39. Richardson, G. (2010); Summary of Waste to Wheels, U.S. Department of Energy, Clean Cities Program, by Energy Vision.

40. Schieder, D., Schneider, R. and Bischof, F., (2000); Thermal Hydrolysis (TDH) as a Pre-Treatment Method for the Digestion of Organic Waste. Water Science and Technology, 41 (3), 181-187.

41. SEAI (2014); The Process and Techniques of Anaerobic Digestion–Digestion Inhibitors http://www.seai.ie/Renewables/Bioenergy/Bioenergy_Technologies/Anaerobic_Digestion/The_Process_and_Techniques_of_Anaerobic_Digestion/Digestion_Inhibitors.pdf.

42. Serfass, P., Biogas Processing for Utilities, , Executive Director, American Biogas Council (www.americanbiogascouncil.org).

43. Seadi, T. A., Rutz, D., Prassl, H., Köttner, M., Finsterwalder, T., Volk, S., Janssen, R. (2008); Biogas Handbook, http://www.seai.ie/Renewables/Bioenergy/Bioenergy_Technologies/Anaerobic_Digestion/Introduction_to_Anaerobic_Digestion/

44. Shahriari, H., Warith, M., Hamoda, M., Kennedy, K.J. (2011); Anaerobic Digestion of Organic Fraction of Municipal Solid Waste Combining Two Pretreatment Modalities, High Temperature Microwave and Hydrogen Peroxide. Journal of Waste Management, 32 (2012) 41–52.

45. Steffen, R.; Szolar, O. and Braun, R. (1998); Feedstocks for Anaerobic Digestion, http://www.adnett.org/dl_feedstocks.pdf.

46. Sun, Y., Cheng, J. (2002); Hydrolysis of Lignocellulosic Materials for Ethanol Production: A Review, Bioresource Technol., 83:1–11.

47. Taherzadeh, M. And Karimi, K. (2008); Pretreatment of Lignocellulosic Wastes to Improve Ethanol and Biogas Production: A Review, Int J Mol Sci. Sep 2008; 9(9): 1621–1651.

48. Tiehm, A., Nickel, K. and Neis, U. (1997); The Use of Ultrasound to Accelerate the Anaerobic Digestion of Sewage Sludge, Water Science and Technology, Vol-36, No.11, 121–128.

49. Valo, A., Carrère, H. and Delgenès, J. P. (2004); Thermal, Chemical and Thermo-Chemical Pre-Treatment of Waste Activated Sludge for Anaerobic Digestion, Journal of Chemical Technology and Biotechnology, Volume 79, Issue 11, pages 1197–1203.

50. Walker, M., Iyer, K., Heaven, S., and Banks, C. J. (2011); Ammonia Removal in Anaerobic Digestion By Biogas Stripping: An Evaluation of Process Alternatives Using a First Order

Rate Model Based on Experimental Findings, Chemical Engineering Journal, 178(15), 138–145.

51. Water Environment Federation (2004); High Performance Anaerobic Digestion White Paper.

52. Woodard, S.E. and Wukasch, R.F. (1994); A Hydrolysis/ Thickening/Filtration Process for the Treatment of Waste Activated Sludge. Water Science and Technology, Vol-30, No. 3, 29–38.

53. WRAP (2012); Enhancement and Treatment of Digestates From Anaerobic Digestion Written by Pell Frischmann Consultants Ltd.

54. Yasui, H. and Shibata, M. (1994); An Innovative Approach to Reduce Excess Sludge Production In the Activated Sludge Process, Water Science and Technology, Vol-30, No. 9, 11–20.

55. Yirong, C., Banks, C. J. and Heaven, S. (2013); Comparison of Mesophilic and Thermophilic Anaerobic Digestion of Food Waste, www.redbiogas.cl/wordpress/wp-content/uploads/2013/07/IWA-12154.pdf.

www.ingramcontent.com/pod-product-compliance
Lightning Source LLC
Chambersburg PA
CBHW041721210326
41598CB00007B/731